Robot

Mere Machine to Transcendent Mind

Hans Moravec

New York Oxford
OXFORD UNIVERSITY PRESS
1999

Oxford University Press

Oxford New York
Athens Auckland Bangkok Bogotá Buenos Aires
Calcutta Cape Town Chennai Dar es Salaam Delhi
Florence Hong Kong Istanbul Karachi Kuala Lumpur
Madrid Melbourne Mexico City Mumbai Nairobi Paris
São Paulo Singapore Taipei Tokyo Toronto Warsaw

and associated companies in

Berlin Ibadan

Published by Oxford University Press, Inc.
198 Madison Avenue, New York, New York, 10016

Library of Congress Cataloging-in-Publication Data
Moravec, Hans P.
Robot: mere machine to transcendent mind /
by Hans Moravec.
p. cm.
ISBN 0–19–511630–5
1. Robotics I. Title
TJ211.M655 1999 303.48'34'0112—dc21 97–47328

1 3 5 7 9 8 6 4 2

Printed in the United States of America on acid free paper

Contents

	Preface	vii
1	Escape Velocity	1
2	Caution! Robot Vehicle!	15
3	Power and Presence	51
4	Universal Robots	91
5	The Age of Robots	127
6	The Age of Mind	163
7	Mind Fire	191
	Notes	213
	Acknowledgments	219
	Index	221

Preface

This book has been brewing for nearly fifty years, since preschool adventures with a mechanical construction set implanted the consuming notion that inanimate parts could be assembled into animate beings. The brew bubbled over in an article in 1978, a book in 1988, and this work in 1998.

In 1978 I was at the Stanford Artificial Intelligence Laboratory finishing a thesis on a seeing mobile robot, and disagreeing with SAIL's pioneering founder John McCarthy, whose focus was computer reasoning. Reasoning programs had bested humans in specialized areas and John held that existing computers lacked only the right programs to be fully intelligent. Computer vision convinced me otherwise. Each robot's-eye glimpse results in a million-point mosaic. Touching every point took our computer seconds, finding a few extended patterns consumed minutes, and full stereoscopic matching of the view from two eyes needed hours. Human vision does vastly more every tenth of a second.

No machine in 1978 even approached everyday human sensory, motor, or reasoning skills. Worse, progress toward those goals seemed to have almost stalled. Reacting against a pervasive pessimism, I argued in an evolving tract that humanlike performance needed millionfold greater computer power, but might be attained in a crash program in as little as ten years by interconnecting millions of then-new microprocessors. *Analog* science fiction magazine bought a version of this polemic, titled "Today's Computers, Intelligent Machines and Our Future," in 1978.

The *Analog* article implausibly called for someone to invest billions of dollars in computer hardware to possibly produce one humanlike machine. Updates of the article were more sedate, offering human competence in multimillion-dollar computers in twenty years. But in AI research, power per computer was falling! Mainframe computers owned by institutions were replaced by minicomputers owned by groups, then by workstations and personal computers. Still, the price/performance ratio was improving, and the decline in individual machine power had to bottom eventually. I resolved to redo my estimates more carefully, and expand the essay into a book. In 1988's *Mind Children: The Future of Robot and Human Intelligence*, humanlike performance was rescheduled for ten-thousand-dollar personal computers in forty years.

By 1990 personal computers outpowered 1978 mainframes, and formerly intractable problems began to find solutions. Home computers soon recognized printed and spoken words, and experimental robots cruised hallways and highways. With a firmer launch ramp than *Mind Children*, this book projects humanlike competence in thousand-dollar computers in forty years. A slight rise in the estimated difficulty has been partially offset by faster growth in computer power.

Mind Children's near-term projections have held up pretty well, with one big exception. There are still no mobile utility robots to help us around the house. In 1988 several small companies had developed security, cleaning, and transport robots that navigated prearranged paths by sensing walls and special markers, somewhat like insects. I had hoped those would lead to more advanced (and more marketable) robots that perceived their surroundings and roamed freely. Alas, the first machines did not attract customers, and the companies fizzled. There are perceiving robots now, but only in research settings. But, this time *for sure*! In line with refurbished predictions for utility robots set forth in Chapter 4, I am pursuing a business plan to equip existing factory vehicles with enough spatial awareness to work in unmapped areas by about 2000, expanding to mass-market machines like small home vacuum robots by 2005, then more capable multipurpose "universal" robots by 2010. It is becoming a race: many other robot ventures are afoot.

Chapter 4 anticipates four generations of universal robot, each spanning a decade. The first has a lizardlike spatial sense, the second adds mouselike adaptability, the third monkeylike imagination, and the fourth humanlike reasoning. The uniform schedule comes from matching each prototype animal's brain to steadily rising computer power. But it took three hundred million years for our ancestors to evolve from tiny ancestral chordates (of insect-like complexity, like robots now) into something lizardlike, an additional one hundred and fifty million to become mouselike, only seventy million more to be monkeylike, and just another fifty million to become human. My time scale is probably too bold in the short run and too timid in the long run, a typical problem with technological forecasts. A future book (dated, say, 2008!) will surely contain a course-corrected projection.

Until then, look for new developments, commentary, references, and color illustrations on my world-wide-web page containing the search word "`hpm98book`," found, for now, at the location:

`http://www.frc.ri.cmu.edu/~hpm/book98/`

1

Escape Velocity

Progressive change sculpted our universe and our societies, but only very recently has human culture seen beyond the short cycles of day and night, summer and winter, birth and death, to recognize it.[1] No sooner was universal change noted in the traces of history than its accelerating pace became discernible in a single lifetime. By almost any measure—energy, information, speed, distance, temperature, variety—the developed world is growing more capable and complex faster than ever before—a statement that has been true for at least half a millennium, and mostly true since the agricultural revolution and the invention of writing over five thousand years ago.

Many of the products of this accelerating process—written language, city-states, and automation, for instance—sped it further. Today the pace strains the limits of human adaptability: the lessons of a technical education are often obsolete before the education is complete. Nevertheless, the acceleration continues, as machines take over where humans falter. In the 1970s photographic patterns for manufacturing integrated circuits with dozens or hundreds of components were designed and drafted by hand, on plastic sheets. Today's computer chips contain tens of millions of components, placed by design programs running on older computers. Not only did one generation of machines make possible the next, they enabled it to appear in less than a year, compared to an average of three years for purely human designs.

Self-accelerated computer evolution affects other technical fields, and almost every design engineer's work is being amplified and accelerated by computer workstations and communications. The many parts of the new Boeing 777 aircraft, to pick a major example, were designed in parallel, by distantly separated engineering teams, with powerful three-dimensional modeling programs. Subassembly designs were checked for compatibility by programs that put together the aircraft in simulation, detecting major and minor mechanical, electrical, control, and aerodynamic problems while they were easy to correct, long before a physical prototype was built. The result was an aircraft of unprecedented complexity brought into existence in half the time of previous models. In the same way, chemists and biologists are replacing years of wet lab work with weeks of molecular simulation. Architects quadrupled their business capacity by replacing drafting boards and manual bookkeeping with computer workstations in the early 1990s.

Vanishing Verities

"Things tossed up come down" is an early theory of gravity, demonstrably true in everyday life, unquestioned for millennia, until Newton developed a new theory of gravity that gave stable orbits to sufficiently fast satellites, and let slightly faster projectiles escape to infinity. "Wood rubbed warm cools down" may have been a truism for our distant ancestors, until one of them rubbed hard enough to achieve ignition temperature, whereupon the wood flamed hotter than ever on its own. "Machines break down" is a demonstrable truth of industrial society, but as machines increasingly design, diagnose, and repair themselves, it too will be suddenly invalidated. Once given "escape velocity," machines more capable than any we know will, without further help from us, grow more capable still, learning from the world, as we did in our biological and cultural evolution. The wood is already smoldering.

Like passengers in a rising elevator, those riding a developmental curve may be unaware of the altitude already reached—until a passing window shows a glimpse of the ground. In 1930 an Australian gold-prospecting party flew into a supposedly uninhabited area deep in the New Guinea highlands and encountered a human culture separated fifty thousand years from their own. The naked inhabitants, some with stone spears, were driven into paroxysms of confusion and religious fear and awe by the giant roaring silver birds that alighted near their mud-thatch villages to release droopy-skinned white men without genitals who, among too many wonders, captured their souls in small black boxes labeled Kodak.[2]

In 1991 Davi Kopenawa took the giant step of being the first to leave the jungle to speak for his people, the Amazonian Yanomami. The Yanomami, with a population of about twenty thousand the largest remaining stone-age tribe, were isolated from the rest of the world for ten thousand years until this century, when missionaries, anthropologists, and, more recently, highway workers and gold miners, began to invade their homeland. Accompanied by a translator, wearing his only possessions, sneakers, jeans, and a sweater given to him for the trip, he visited New York, Washington, and Pittsburgh, to beg to be left alone: foreign diseases, especially malaria, had killed one-fifth of the Brazilian Yanomami in five years.

What he saw in the cities horrified him: crazy ant-people crawling in sky-high huts thinking about cars, money, and possessions instead of relatives and nature. In a zoo he identified with the listless animals among plastic plants, steel vines, and bad air. "If I had to live in your cities for a month, I'd die. There's no forest here."

Kopenawa has a point. The world we inhabit is radically different, culturally and physically, from the one to which we adapted biologically. We were shaped during the last two million years by an ongoing Ice Age—a time of continuous climatic change, as every few tens of thousands of years glaciers advanced and retreated over most of the earth (the current warm spell is but an interglacial period). Such variability favors high adaptability, by making life untenable for the rigidly optimized. In our species the adaptability took the form of a hypertrophied brain and an extended childhood, supporting an extreme cultural plasticity, along with an ever more expressive language to rapidly pass on adopted behaviors: as we grow to puberty we can learn equally well to be fur-clad arctic hunters, robed desert nomads, or naked equatorial gatherers. For almost all of human history, as still in Kopenawa's world, cultural inheritance played a straightforward supporting role: providing the *how* for the basic needs of life. But somewhere, about five thousand years ago in our cultural history, the relationship between biology and culture began to alter radically.

The Cultural Revolution

Culture lets us rapidly accommodate to environmental changes because it is a medium for a new kind of evolution. Collections of rules for behavior (memes, to use a term invented by Richard Dawkins[3]) pass from generation to generation, mutating and competing with alternatives, just as biological genes do—only much more quickly. A biological trait requires generations of selective replication to become established in a population, but a cultural practice can be altered, and spread through an entire tribe, many times in a single human lifetime. After hundreds of thousands of years of slow cultural meander, our ancestors stumbled into a set of behaviors that catalyzed the creation of ever more behaviors and memories, and physical implements to support them: a self-accelerating cycle that is reaching escape velocity today. What exactly sparked the tinder, apart from simple accumulation of useful skills, is a fascinating question. A baby boom or forced migration in an improving climate may have led to shortages in hunting and gathering resources, forcing would-be survivors into agricultural life, and eventually the first agricultural civilizations, in the Near East and China ten millennia ago.

For millions of years, primates, our ancestors included, have lived in tribes. Among primates (as in canine packs, but unlike herd animals) individuals know one other personally, and maintain long-term one-on-one relationships, involving dominance, submission, friendship, enmity, debts, grudges, and intrigues—the stuff of soap opera. Complex socialization gives the tribe great abilities. In critical circumstances, individuals know who to trust with what tasks. But remembering many things about many individuals ought to take storage space in the brain. Robin Dunbar[4] has indeed found a tightly correlated linear relation between brain size and troop size in monkeys and apes—macaque monkeys, for instance, form bands of about fifteen, while larger-brained chimpanzees and gorillas live in tribes of thirty to forty members. This soap-opera connection likely drove the evolution towards large brains in pri-

inhabitants, some with stone spears, were driven into paroxysms of confusion and religious fear and awe by the giant roaring silver birds that alighted near their mud-thatch villages to release droopy-skinned white men without genitals who, among too many wonders, captured their souls in small black boxes labeled Kodak.[2]

In 1991 Davi Kopenawa took the giant step of being the first to leave the jungle to speak for his people, the Amazonian Yanomami. The Yanomami, with a population of about twenty thousand the largest remaining stone-age tribe, were isolated from the rest of the world for ten thousand years until this century, when missionaries, anthropologists, and, more recently, highway workers and gold miners, began to invade their homeland. Accompanied by a translator, wearing his only possessions, sneakers, jeans, and a sweater given to him for the trip, he visited New York, Washington, and Pittsburgh, to beg to be left alone: foreign diseases, especially malaria, had killed one-fifth of the Brazilian Yanomami in five years.

What he saw in the cities horrified him: crazy ant-people crawling in sky-high huts thinking about cars, money, and possessions instead of relatives and nature. In a zoo he identified with the listless animals among plastic plants, steel vines, and bad air. "If I had to live in your cities for a month, I'd die. There's no forest here."

Kopenawa has a point. The world we inhabit is radically different, culturally and physically, from the one to which we adapted biologically. We were shaped during the last two million years by an ongoing Ice Age—a time of continuous climatic change, as every few tens of thousands of years glaciers advanced and retreated over most of the earth (the current warm spell is but an interglacial period). Such variability favors high adaptability, by making life untenable for the rigidly optimized. In our species the adaptability took the form of a hypertrophied brain and an extended childhood, supporting an extreme cultural plasticity, along with an ever more expressive language to rapidly pass on adopted behaviors: as we grow to puberty we can learn equally well to be fur-clad arctic hunters, robed desert nomads, or naked equatorial gatherers. For almost all of human history, as still in Kopenawa's world, cultural inheritance played a straightforward supporting role: providing the *how* for the basic needs of life. But somewhere, about five thousand years ago in our cultural history, the relationship between biology and culture began to alter radically.

The Cultural Revolution

Culture lets us rapidly accommodate to environmental changes because it is a medium for a new kind of evolution. Collections of rules for behavior (memes, to use a term invented by Richard Dawkins[3]) pass from generation to generation, mutating and competing with alternatives, just as biological genes do—only much more quickly. A biological trait requires generations of selective replication to become established in a population, but a cultural practice can be altered, and spread through an entire tribe, many times in a single human lifetime. After hundreds of thousands of years of slow cultural meander, our ancestors stumbled into a set of behaviors that catalyzed the creation of ever more behaviors and memories, and physical implements to support them: a self-accelerating cycle that is reaching escape velocity today. What exactly sparked the tinder, apart from simple accumulation of useful skills, is a fascinating question. A baby boom or forced migration in an improving climate may have led to shortages in hunting and gathering resources, forcing would-be survivors into agricultural life, and eventually the first agricultural civilizations, in the Near East and China ten millennia ago.

For millions of years, primates, our ancestors included, have lived in tribes. Among primates (as in canine packs, but unlike herd animals) individuals know one other personally, and maintain long-term one-on-one relationships, involving dominance, submission, friendship, enmity, debts, grudges, and intrigues—the stuff of soap opera. Complex socialization gives the tribe great abilities. In critical circumstances, individuals know who to trust with what tasks. But remembering many things about many individuals ought to take storage space in the brain. Robin Dunbar[4] has indeed found a tightly correlated linear relation between brain size and troop size in monkeys and apes—macaque monkeys, for instance, form bands of about fifteen, while larger-brained chimpanzees and gorillas live in tribes of thirty to forty members. This soap-opera connection likely drove the evolution towards large brains in primates, since tribes compete for food and shelter-providing territory, and a coordinated larger group is likely to beat out a smaller one, giving large tribes, and thus large brains, an advantage. Dunbar extrapolates the primate group/brain ratio curve to human brain size, and finds our natural tribes should have about two hundred individuals. In fact, this is just the maximum size of self-contained nonhierarchical human groups: Yanomami Indian villages and gypsy bands, for instance, and perhaps hippie communes. Modern society's overlapping webs of individual acquaintances muddle but don't eliminate our tribal limitations, evidenced in ubiquitous anecdotes. My wife, involved in many church organizations, notes that growing churches have major crises of identity when their membership reaches about two hundred. The computer science department at Carnegie Mellon University was known for its cooperative, "family" atmosphere in the 1970s, when it numbered about a hundred. It grew rapidly in the 1980s, and in the 1990s the over six hundred members of the School of Computer Science are divided into several departments and projects that are strangers to one another.

The agricultural civilizations were able to grow far beyond village size because of a series of social inventions, among them institutional roles like King, Soldier, Priest, Merchant, Tax Collector, and Peasant, clearly marked by costumery, ceremony, and standard rituals, substituting for the impossible task of remembering thousands of individual relationships. New solutions bring new problems. Cheaters in villages are easily recognized and punished but find many opportunities and hiding places in the anonymity of large society. Enforcement institutions—Moral authorities, Lawmakers, Police, and Criminal labels—partially countered the breakdown of cooperation. The problem of keeping track of who owes what to whom, a matter of memory in a village but a criminal opportunity in a city, encouraged the invention of recordkeeping: tokens, tally marks, a number system, and eventually writing. The new social functions involved complicated procedures unlike those of tribal life, many thus slow and difficult to learn. Enter extended formal training periods, eventually Teachers and Schools.

Like villages, civilizations compete with one another for resources and may gain advantage from institutions that foster

Community life and chipped silicon

For over a million years we evolved biologically into circumstances resembling the images on the left. The bizarre variants on the right evolved culturally, much too fast for our biology to have kept up. It is a credit to the flexibility of our ancestral design that many of us manage to squeeze into the strange new molds. It is no surprise if we find the fit uncomfortable—or that some of us never manage it at all.

innovations—and incidentally put cultural evolution into higher gear. Agriculture benefits from precise knowledge of the season, and thus of celestial cycles, and military and civil projects go better with professional thinkers and builders on the job, so the positions of Astronomer/Astrologer, Philosopher/Magician, and Engineer/Artisan become part of the picture. The innovations of professional thinkers, transmitted by increasingly effective written language over huge distances and times, accelerate innovation itself. The result is a process far, far faster than biological evolution that produces ever more elaborate places for humans to live, ever swifter ways for them to move and to communicate, ever larger storehouses of previous thought, ever more territory occupied, ever more energy controlled. It also produces a world increasingly unlike the villages, fixed and nomadic, in which human behavior evolved, and so makes ever greater demands on our adaptability.

Strange Ducks, Out of Water

Today, as our machines approach human competence across the board, our stone-age biology and our information-age lives grow ever more mismatched. Work in the developed countries has become increasingly specialized and esoteric, and it now often takes a graduate degree, representing half a working lifetime of sustained learning, to master the necessary unnatural skills. As societal roles become yet more complex, specialized, and far removed from our inborn predispositions, they require increasing years of rehearsal to master, while providing fewer visceral rewards. The essential functions of a technical society elude the understanding of an increasing fraction of the population. Even the most successful individuals often find their work boring, difficult, unnatural, and unsatisfying, more like a sustained circus performance than a real life. Caffeine substitutes for natural adrenaline. Those original activities that do remain—eating and child raising, for instance—are often squeezed by the strange new tasks. The mismatch between instinct and necessity induces alienation in the midst of unprecedented physical plenty.

By the standards of our inherited tribal psychology, we *are* sick and crazy. Physically, however, we are healthier and live longer than ever, and we have vastly more options in every sphere of activity. Few city-dwellers would be prepared to adopt the circumscribed life in a stone-age forest village, despite uneasiness with their own. On the contrary, much of the third world is rushing to overcome its physical problems by adopting the patterns of the developed nations (Davi Kopenawa himself is now a regular speaker at ecological meetings around the world, and his stories and commentary are extensively represented on the World Wide Web!). The urbanized, meanwhile, have devised substitutes for some tribal experiences, for instance, churches and other social organizations that bring together village-sized groups with a common sense of purpose, a shared experience, a defining mythology, and uniform behavioral expectations. Others find release in competitive sports (very like tribal wars), outdoor vacations, or even backyard barbecues. Some business trips resemble mammoth-hunting forays but lack the scenery, quiet stalks, and satisfying physical marksmanship—and a golfing weekend fills the void. But, as the pace, diversity, and global geographic interconnectedness of life continues to increase, even such occasional imitations of our ancestors' lifestyles are crowded out and may be becoming less satisfying. The world is rushing away from our ancestral roots ever faster, stretching the limits of both our biological and institutional adaptability.

Some individuals and communities have tried to isolate themselves from the problem. The Pennsylvania Amish live in a perpetual state of early-nineteenth-century rural industrialization. Some cloistered religious orders operate like isolated tribes. Countercultural rural communes of the 1960s and 1970s deliberately resembled villages. Yet, industrialized society's increasing population, accessibility, and competitive vigor in all fields seems to erode such communities, who cannot reasonably, or legally, deny members in need the benefits of modern medicine, inexpensive food, clothing, building materials, useful machinery, and especially empowering, but distracting, education.

There are unhappy voices today calling for a worldwide rollback to an earlier state of affairs. They are outvoted by the demands of billions for food, housing, and civilized comforts. Yet, paradoxically, as our cultural artifacts achieve self-sustaining maturity, they will provide the means to restore humanity and nature to an imitation of the wild past.

Back to the Future

Productivity rose during the Industrial Revolution, as steadily improving machines outperformed and displaced ever more human labor. Simple diffusion, and social innovations like labor unions and profit taxes, widely distributed the consequent wealth. The wealth expressed itself both in increased public and private consumption and increased leisure time. During the last three centuries in the industrialized countries, slave and child labor and hundred-hour factory work weeks have given way to under-forty-hour weeks and mandatory retirement.

Short-term fluctuations in the trend notwithstanding, as machines assume more—eventually all—of essential production, humans everywhere will be left with the options of the idle rich. Work time is yoked to the strange needs of the enterprise, but idle time can be structured to satisfy hunter-gatherer instincts. The human population will regain the opportunity to organize its life in more natural patterns. A greener planet is a likely result of this ongoing process. As societies industrialize and become wealthy, increased consumption manifests itself in deforestation, pollution, and the like—to a point.

Further wealth reduces the manifestations of industry, by making the luxury of a greener environment affordable. Advancing technology widens the options from which individuals sculpt first their personal lives, but then also their communal world. The developed countries of America, Asia, and Europe began their green return in recent decades, as per-capita annual income grew beyond about $15,000. Many developing countries are just reaching this turnaround point. Advanced robots will reinforce the trend indirectly, by tremendously accelerating technological evolution and, for instance, allowing extreme processes to be moved to outer space. They will contribute directly by substituting for energy- and chemical-intensive industrial separation and shaping processes. A robot population far exceeding the human one will achieve the same end much more efficiently by tirelessly sorting and rearrang-

Up, then down
Growing knowledge and wealth increase the options available to individuals. Personal needs and wants outweigh communal concerns, but eventually even communal goods become affordable. Despite increasing population and personal consumption, U.S. environmental pollution has mostly improved since 1970, when per capita income exceeded $15,000 in 1998 currency. A consensus of opinion set the direction, technology and wealth provided the means. (Source: U.S. Statistical Abstract of the United States, 1996, *CDROM, U.S. Census Bureau, 1997, esp. table 374, spreadsheet version. Supplemental data from* Historical Statistics of the United States: Colonial Times to 1970, *U.S. Census Bureau, 1976.)*

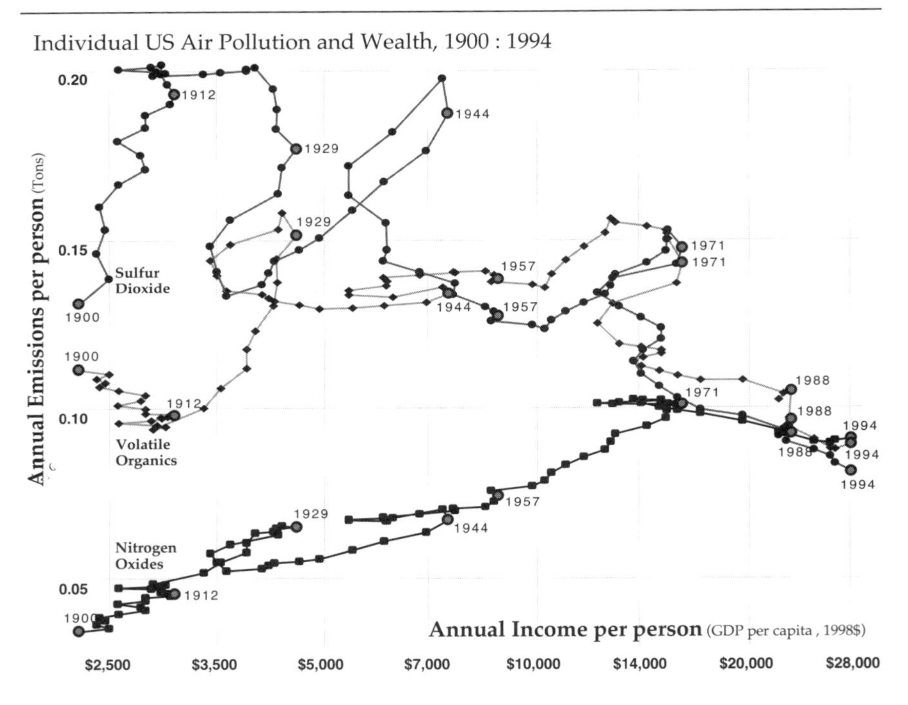

ing matter on a tiny scale with myriads of microscopic fingers.

Any choice has consequences: by comfortably retreating to its roots, biological humanity will leave the uncomfortable, uncharted but grand opportunities to others.

No less than today's organizations, fully automated companies will compete with one another not only in routine manufacture and distribution, but also in planning, development, and research. To robots built for it, outer space will offer unprecedented energy, materials, room, and perhaps freedom from taxation

Up, then down (cont)

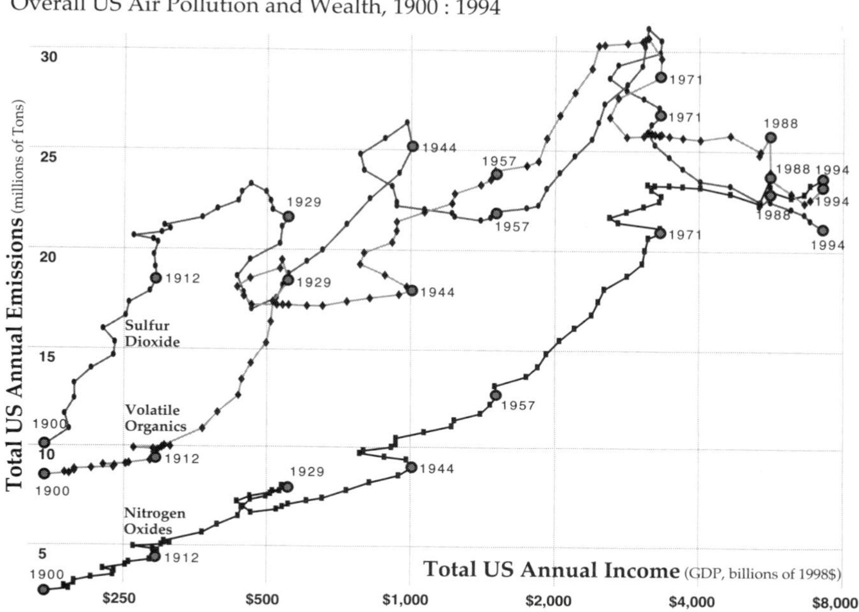

for these activities—a tremendous competitive advantage. Sooner rather than later, automated industry will grow away from earth. The space industries will continuously devise their own improvements, gaining rapidly in size, efficiency, diversity, and intelligence. The earthbound "consumer outlet" parts of the operation, while not shrinking in absolute size, will represent an ever-decreasing fraction of the whole. Old earth will become insignificant on the ever-grander scale of earth-spawned activity.

Robot industries will start as conversions of existing enterprises, retaining their institutional, legal, and competitive structures. But then they will explore and exploit expanding non-traditional options, some very unhuman. Our artificial progeny will grow away from and beyond us, both in physical distance and structure, and in similarity of thought and motive. In time their activities may become incompatible with old earth's continued existence. Even so, it is likely that we, the historical root of their transcendence, will be preserved in some form—though, to us, the form may seem extremely strange. Just possibly, human personalities could participate in some way in the mainstream of this future activity, either under the wings of superintelligent hosts, or by being transformed into a compatible form—surely becoming very unhuman in the process.

There is an analogy between the evolution of the first living organisms from simpler chemical processes several billion years ago and the development of technical civilization from human manipulative and learning skills. Technical civilization, and the human minds that support it, are the first feeble stirrings of a radically new form of existence, one as different from life as life is from simple chemistry. Call the new arrangement Mind. Unlike Life alone, which learns from its past, but is blind to its future, Mind can choose among alternatives to imperfectly select its own destiny—even to amplify that very ability.

Mind Fire

Chapter 2 reviews the state of the robot art, like a baby poised for sudden growth. The following chapters mix predictions with suggested actions. One of the lessons of chaos theory is that sensitive systems are impossible to predict but often easy to control. Under that model, the future *can* sometimes be predicted, if one steadily nudges events toward the prediction! Believable and physically possible predictions can themselves nudge, by inspiring work. When such proactive predictions miss, it is often because they overlook even more potent possibilities, rather than because they are unachievable.

In the thirteenth century Roger Bacon imagined high-speed worldwide travel—via seven-league boots, rather than flying conveyances. In the sixteenth century Leonardo da Vinci designed aircraft—powered by human muscle, rather than combustion engines. In the nineteenth century Jules Verne anticipated submarine warfare—against wooden sailing vessels, rather than armored battle fleets guarded by electronic senses and aircraft. Shortly thereafter, H. G. Wells anticipated a world of the distant future with humanity radically transformed—by Darwinian evolution, not directed engineering. Science fiction of the early twentieth century, inspired by the theories, inventions, and speculations of rocket pioneers like Konstantin Tsiolkovsky, Robert Goddard, and Hermann Oberth, is filled with spacecraft—guided by slide-rule-wielding human navigators, not digital computers (telephone, radio, and computers, in particular, and their dramatic applications, seem to have taken prognosticators by surprise). There are no large fleets of dirigible airships ferrying transatlantic passengers; faster and more manageable heavier-than-air craft displaced them.

Barring cataclysms, I consider the development of intelligent machines a near-term inevitability. Chapters 3 and 4 offer a scenario. Like airplanes, but unlike spaceships or radio, machine intelligences will be direct imitations of something already existing biologically. Every technical step toward intelligent robots has a rough evolutionary counterpart, and each is likely to benefit its creators, manufacturers, and users. Each advance will provide intellectual rewards, competitive advantages, and increased wealth and options of all kinds. Each can make the world a nicer place to live. At the same time, by performing better and cheaper, the robots will displace humans from essential roles. Rather quickly, they could displace us from existence. I'm not as alarmed as many by the latter possibility, since I consider these future machines our progeny, "mind children" built in our image and likeness, ourselves in more potent form. Like biological children of previous generations, they will embody humanity's best chance for a long-term future. It behooves us to give them every advantage and to bow out when we can no longer contribute.

But, as also with biological children, we can probably arrange for a comfortable retirement before we fade away. Some biological children can be convinced to care for elderly parents. Similarly, "tame" superintelligences could be created and induced to protect and support us, for a while. Such relationships require advance planning and diligent maintenance. Chapter 5 offers suggestions.

It is the "wild" intelligences, however, those beyond our constraints, to whom the future belongs. The available tools for peeking into that strange future—extrapolation, analogy, abstraction, and reason—are, of course, totally inadequate. Yet, even they suggest surreal happenings. Chapter 6's robots sweep into space in a wave of colonization, but their wake converts everything into

Cart under SAIL
As both robot vehicle and road hazard, the dimly aware Stanford Cart traverses the driveway of the Stanford Artificial Intelligence Laboratory in a composite image recreating the situation in the 1970s.

cessed and the robot controlled by the AI Lab's room-sized computers, and then the vehicle truly was an autonomous mobile robot. In 1971, in the Ph.D. work of a student named Rod Schmidt, the Cart, in a kind of robotic sobriety test, was programmed to slowly follow a white line. In my own studies, in 1975 it drove in straight lines by keeping its eye on a skyline of trees, and in 1979 peered and planned its way through obstacle courses, crossing a thirty-meter room in the astonishing time of five hours, getting lost about one crossing in four.

Embarrassingly subhuman performance has been typical of robotics, to the chagrin of its enthusiasts. A 1969 essay by John McCarthy, *Computer-Controlled Cars*, suggested that a medium computer of the day could drive a seeing robot car evolved from the Stanford Cart in normal traffic. More conservative visionaries thought automated traffic might soon travel on special highway lanes marked by buried signal-emitting wires. Three decades later, computers grown a hundred thousand times as powerful still can't be trusted to drive vehicles unattended, except for a few people movers that run on sensor-studded tracks.

Why was the intuition of some very clever people so far off the mark? Distances can be deceiving when one sees a territory from only one special vantage point. Distance, in this metaphor, is the absolute difficulty of accomplishing various mental, sensory, and motor feats. The special vantage point is our large brain, sculpted over hundreds of millions of years. During that time our animal ancestors were selected for their proficiency in obtaining the essentials of life and reproduction, and winning escalating arms races with competitors and parasites. In contrast, writing, arithmetic, and logical reasoning are recent cultural artifices. We find sensing and moving natural and easy, while "brain work" is tedious and difficult. An effortless glance reveals the pieces on a chessboard, but it takes long, hard thought to choose good chess moves.

A decades-long effort to build machines that sense, act, and think has established a very different viewpoint. Robot sensing and moving, no less than thinking, is built up of large numbers of simple steps. It has become clear that reliably locating chess pieces in camera images demands thousands of times as many steps as planning good chess moves.

The serious possibility of machines behaving like animals and humans arose from the technology of World War II. On the one hand, servomechanisms, in sensors and weapons, allowed electronically driven motors to follow precise motions and respond to subtle sensory cues. On the other, digital computers, breaking codes, calculating artillery tables, and simulating atomic explosions, executed enormously long arithmetical and logical procedures with superhuman speed and accuracy.

Cybernetic Creatures

Norbert Wiener, a mathematician who had developed the theory for complex predictive bombsights and anti-aircraft gun directors, attracted a small community of psychologists, biologists, engineers, and scientists of other disciplines, mostly in America and Britain, into a new field he called *cybernetics*—the science of control and communication in the animal and the machine.[5] Under that banner, researchers organized shoeboxes of electronics into artificial nervous systems that could learn to recognize simple patterns, and turtle-like robots that avoided obstacles and sought out lights.

1950—Dr. W. Grey Walter operates on one of his light-seeking tortoises
When "Elsie" is fully assembled, the bump-sensing switch at the top of the mechanism supports a protective shell, seen in the background.

In England from 1948 to 1951, W. Grey Walter, a British psychologist, built a half-dozen electronic tortoises with subminiature radio-tube brains, rotating phototube eyes, and contact-switch feelers.[6] They could avoid trouble while wandering about and return to their carrying hutch when its light beckoned. In groups they exhibited unexpected social behavior (dancing) by responding to one another's control lights and touches.

In the early 1960s the availability of transistors, much smaller, cheaper and less power consuming than tubes, allowed a group of brain researchers at the Johns Hopkins University Applied Physics Laboratory (APL) to build more complex robots. The "Hopkins Beast" wandered the halls of APL, keeping itself on the midline

1964—Two early versions of the "Hopkins Beast" feeding at a wall outlet
These machines wandered hallways guided by sonar and found sockets by feeling along walls. In subsequent work, the larger unit was given another layer containing a photocell circuit that allowed it to find contrasting wall plugs optically from a distance.

by side-directed sonar. When its batteries ran low, a specialized photocell eye searched the white-painted cinder-block walls for the distinctive black cover plate of wall outlets, where the robot would plug in to feed. The Beast inspired several imitators. Some used television cameras instead of photocells, and were controlled by assemblies of transistor logic gates, like those that can now be found, in the millions, in the integrated circuits of every computer. Some added new motions such as "Shake to untangle recharging arm" to the repertoire of basic actions.

Cybernetics' early results were intriguing enough to sustain two decades of research, but the field stumbled in the late 1960s when its methods proved unsuccessful when applied to more challenging problems, such as the effort to build practical reading machines.

Artificial Intellects

Electronic computers fueled an entirely different approach to thinking machines. In 1936 Alan Turing invented the mathematical concept of a universal computer that could be programmed to imitate the action of any other information processor; he used it to show that unsolvable problems exist. During World War II, in deepest British secrecy, he put his ideas into practice in the Colossus electronic computers. Those machines broke the German U-boat codes, allowing allied shipping to evade the wolf packs and win the war. After the war, he speculated that the imitation repertoire of a universal computer included all the functions of the human brain. Although his wartime accomplishments remained a state secret until 1974, his speculations on intelligent machinery became public in a series of debates on BBC radio in 1950 and in a following article.

The themes sounded on both sides in those debates still echo, though Turing died in 1953. In the United States, John von Neumann, who had developed mathematics and computers used in the design of atom bombs, was also speculating on artificial imitations of life and thought. His investigation, too, was cut short by death in 1956. But the question was in the air: with clever programming, might not the great capacity of the new electronic "giant brains" be harnessed to perform mental tasks beyond the mere rote?

The first computers seemed to be locomotives of thought. They were as big as locomotives, and performed feats equally awesome, outpulling hundreds of thousands of straining mathematicians on arithmetical calculations. Around 1955 a handful of enthusiastic young academics took up the challenge of extending their repertoire. The field was named *Artificial Intelligence* by John McCarthy at their first conference, at Dartmouth College in 1956. At the Dartmouth conference, Allen Newell, Herbert Simon, and George Shaw showed off their "Logic Theorist," the first working program of the new field. Starting with the axioms of number theory in Russell and Whitehead's famous (and mind-numbing) tome *Principia Mathematica,* the Logic Theorist was able to apply the rules of deductive inference to prove many of the theorems. Disturbingly, while the JOHNNIAC computer, on which Logic Theorist resided, could superhumanly do ten thousand numerical calculations per second, the program overall proved theorems no faster or better than a college freshman recently taught the subject. The program encoded symbols as numbers, needed hundreds of calculations to apply one rule of inference to one logical sentence, and explored many long chains of inference before finding a proof.

Programs were written during the next decade that proved theorems in geometry, solved calculus problems, and played good games of checkers and chess. Despite a number of clever mathematical and programming innovations, these programs remained at freshman competence. It also became apparent that each program could not be expanded beyond its initial narrow expertise. In particular, there seemed no way to give the programs "common sense" about everyday matters.

The Hard Easy

The discrepancy between the giant brain and the mental midget image of computers became worse in the late 1960s and early 1970s. Marvin Minsky's research group at the Massachusetts Institute of Technology and John McCarthy's at Stanford University connected television cameras and robot arms to their computers so "thinking" programs could begin to collect information directly from the real world. The early results were like a cold shower. While the pure reasoning programs did their jobs about as well and fast as college freshmen, the best robot-control programs, besides being more difficult to write, took hours to find and pick up a few blocks on a tabletop, and often failed completely, performing much worse than a six-month-old child. The disparity between programs that calculate, programs that reason, and programs that interact with the physical world holds to this day. All three have improved over the decades, buoyed by a more than millionfold increase in computer power in the fifty years since the war, but robots are still put to shame by the behavioral competence of infants or small animals. Computers have played grandmaster chess since the 1980s, and IBM's "Deep Blue" machine defeated Garry Kasparov, probably the best human player ever, in a May 1997 match. But Deep Blue needed a human assistant to see and physically manipulate the pieces. No robot existing could have done it in the wide range of circumstances Kasparov, or any child, finds trivial.

For machines, calculating is much easier than reasoning, and reasoning much easier than perceiving and acting. Why is this order of difficulty opposite for humans? For a billion years every one of our ancestors achieved that status by winning a competition for the essentials of life in a hostile world, often by a lifetime of sensing and moving more effectively than the competition. That escalating Darwinian elimination tournament bequeathed us brains spectacularly organized for perception and action—an excellence often overlooked because it is now so commonplace. On the other hand, deep rational thought, as in chess, is a newly acquired ability, perhaps less than one hundred thousand years old. The parts of our brain devoted to it are not so well organized, and, in an absolute sense, we're not very good at it—but, until computers arrived, we had no competition to show us up. Arithmetical calculation lies even further down the spectrum of human proficiency: it became a needed skill—for specialists—only in recent millennia, as civilizations accumulated large populations and pools of property. It is a tribute to our general-purpose perceptual, manipulative, and language skills, and our luck in finding compatible representations of numbers, that we can do arithmetic at all.

By a calculation later in this chapter, it takes a million times the power of a mid-1990s home computer to match a human brain doing what it does best—perception or motor control. Arithmetic lies at the other extreme of human performance. Occasionally a human is found who, through some biological quirk, can perform prodigious feats of mental calculation: some "lightning calculators" can add dozen-digit numbers as fast as they are presented and multiply them in a minute. Yet a home computer can best this performance many millionfold. Between these extremes, in certain tasks of rational thought, like playing chess, proving theorems, or diagnosing infections from lists of symptoms, present computers approximately match human performance.

When the reasoning problem strays beyond narrow logical confines, requiring some kind of visualization, or broad general knowledge, computers again lose out. To this day AI programs exhibit no shred of common sense—a medical diagnosis program, for instance, may prescribe an antibiotic when presented with a description of a broken bicycle because it lacks a model of people, diseases, or bicycles. Yet these programs, on existing computers, would be overwhelmed were they to be bloated with the details of everyday life, since each new fact can interact with the others in an astronomical "combinatorial explosion." [A project begun in 1985 called *Cyc* (from En*cyc*lopedia) at the Microelectronics and Computer Consortium in Austin, Texas, attempted to build just such a common-sense data base. Its founders estimated the final result would contain over one hundred million logic sentences about everyday things and interactions. The project continues today, but with the more modest claim of producing large databases for applications such as image retrieval.]

Machines have a lot of catching up to do. But catching up they are—for most of the century, machine calculation has improved a thousandfold every twenty years. The rate is twice as fast now, and developments already in the research pipeline should sustain it for at least several decades more. In less than fifty years, inexpensive computers will match and exceed—in raw information-processing power—even the well-developed functions of the human brain. But what are the prospects for programming these powerful machines to perceive, intuit, and think like humans?

Cockroach Race

Cybernetics attempted to copy nervous system function by imitating its physical structure. The approach stumbled in the 1960s on the difficulty of constructing all but the simplest artificial nervous systems, then regained vigor in the 1980s under the rubric *neural nets*, as computers became powerful enough to simulate interesting assemblies of neurons. Neural nets are taught to produce desired outputs from given inputs by tweaking the strength of interconnecting links in repeated training steps. In some pattern recognition and motor control tasks where the input/output relation is poorly understood but not too complex, they sometimes provide better performance for less effort than other programming techniques. However, with a maximum of a few thousand neurons, less than an insect's, present simulations cannot orchestrate elaborate behavior.

Achieving intelligent behavior by directly copying the nervous system is circumscribed by another problem. The larger structures and functions of real nervous systems remain mysterious because examining a brain in operating detail is a very slow and tedious process. New instruments, able to scan living nervous systems at high speed in minute detail, may change that in the future.

Artificial Intelligence has successfully imitated the conscious surface of rational thought, and in doing so made evident the vast unplumbed sea of unconscious processes below. Building truly intelligent machines means exploring this ocean, from its rational top to its adaptive bottom. Probably the journey could be completed from either extreme, with growing computer power hosting ever-more-elaborate adaptive systems, or ever-more-comprehensive reasoning programs.

What will get there first, I feel, is a pragmatic broadside in the new field of Robotics, which shamelessly throws together programs that reason, programs that adapt, and powerful mathematical techniques from other parts of computer science to construct systems that see and act in the physical world. I see robotics progress as roughly recapitulating the *evolution* of biological minds, producing a succession of machines whose capabilities resemble those of animals with increasingly complex nervous systems. The effort is guided by observation of the neuronal, structural, and behavioral features of animals and humans, affected by success or failure in the marketplace, and open to any solution to the problems that arise.

Today's best commercial robots are controlled by computers just powerful enough to produce insect-grade behavior. They cost as much as houses, and find only a few profitable niches in society, among them spray-painting and spot-welding cars, assembling electronics, and carting subassemblies from place to place in factories. Only about one hundred thousand exist worldwide. But exotic applications like planetary exploration and the mundane promise of potentially much larger markets for better-performing machines, have encouraged decades of government- and industry-funded research, at modest levels. Progress was frustratingly slow. Finally, in the 1990s, the effort, buoyed by rising computer power, has begun to blossom. Research machines are beginning to navigate well in ordinary indoor and outdoor settings, and commercial programs read text and transcribe speech.

Evolution of full machine intelligence will greatly accelerate, I believe, as mass-produced robots appear in the next decade and slowly broaden in application to become general purpose, *universal* robots. These machines will, at first, be in the physical world what personal computers have been in the world of data—literal-minded followers of prearranged sequences of commands. Gradually they will grow in skill and autonomy—and eventually surpass us in everything. They are the subject of the next chapter. This chapter is about the slow buildup that heralds their arrival.

From the mid-1960s to the mid-1980s the majority of industrial robots were robot arms used in automobile assembly, and most robot researchers concerned themselves with manipulators. Most robot arms are fixed, and they encounter a very limited range of circumstances. In both industrial and research settings, they tend to work best with specialized grippers, sensors, and setups. Specialization is cost effective but antithetical to many of the requirements of universal robots. Fixed robot arms seem to be on a different evolutionary track. Mobile robots, though, seem to be right on track, necessarily flexible in perception and action because they travel through a diverse and surprise-filled world.

Mobile robotics is the youngest segment of the young field of computer-controlled robotics, but since 1985 it has become the largest. One of the few pre-1985 efforts has roots in the Apollo moon landing program. Since about 1965 NASA's Jet Propulsion Laboratory in Pasadena has worked on semiautonomous lunar and planetary rovers. That effort waxed and waned as the prospects of actually launching a rover rose and fell over the decades. It has finally borne fruit in the Sojourner Mars rover, which explored a patch of Mars in 1997, and its smarter successors. Another was Stanford Research Institute's robot Shakey,[7] operated from 1966 to 1972, the first mobile robot to be controlled by a computer. Yet another was the Stanford University Cart project, running from 1968 to 1980, which resulted in two Ph.D. theses, including mine. Several smaller mobile robot research efforts appeared worldwide toward the end of the 1970s, made possible by the appearance of small, relatively cheap computers. I can tell the Cart project's story in some detail.

Cartography

Was it possible to make good on the promises of John McCarthy's *Computer-Controlled Cars* essay? In 1969 the only autonomous computer-controlled wheeled vehicle was a slow-moving 1.5-meter-tall indoor robot, wobbling on a primitive suspension, named Shakey, at Stanford Research Institute or SRI, a contract-research company near Stanford University. Shakey was a child of the first wave of Artificial Intelligence: at its heart was a reasoning program that used theorem-proving methods to contemplate rooms, walls, doors, paths, and large blocks and wedges that could serve as obstacles or objects to be pushed around. Programs to interpret camera and rangefinder data into scene descriptions for the reasoning program, and to cause the robot to actually execute the resulting plans, were considered peripheral and assigned to support engineers. In the early 1970s computer vision was less than ten years old, and almost all was of the "blocks-world" type, initially developed at MIT for finding children's blocks on a tabletop. The SRI team constructed a world to match the means: an array of blank-walled rooms containing a few uniformly painted large blocks and wedges. Shakey was the star of a somewhat misleading 1970 *Life* magazine article, but its most impressive feat—moving a wedge to a block, ascending it, and pushing off a smaller block—was recorded on film piecemeal, requiring multiple takes—and several hours—for each error-prone stage.

McCarthy's challenge was unexplored territory. How could one expect to interpret an image sequence at the many-frames-per-second rate probably necessary for driving, on a 1/2 MIPS (**M**illion-**I**nstruction-**P**er-**S**econd, each instruction causing work similar to adding two eight-digit numbers) machine like SAIL's 1969 computer, a Digital Equipment Corporation PDP-10? A good digitized picture of the road is, by itself, an array of a fraction of a million numbers. To even *touch* each of these pixels (picture elements) with a program took several seconds—and doing anything substantial at least several times longer. How would it be possible to respond swiftly to traffic, obstacles, road signs, and other items that flash through all parts of the image?

Maybe programs could somehow select and work with only the

1970—Shakey the robot reasons about its blocks

Built at Stanford Research Institute, Shakey was remote controlled by a large computer. It hosted a clever reasoning program fed very selective spatial data, derived from weak edge-based processing of camera and laser range measurements. On a very good day it could formulate and execute, over a period of hours, plans involving moving from place to place and pushing blocks to achieve a goal.

essential parts of each image. In 1971 Rod Schmidt used this approach in the first Cart thesis. With Schmidt's program, about two hundred thousand memory-straining bytes of strenuous-to-write but efficient assembly language, the Cart, moving at a very slow walking pace, visually followed a white line on the ground. The program contained a predictor for the future position of the line in the camera scene, based on its past position, and searched about 10% of the next image to find it again. The newfound location served as the next input to both the predictor and a steering calculation. Using about 1/4 MIPS, half the power of the PDP-10, the program could handle one image a second, and enabled the Cart to follow a line for about 15 meters at a time—if the line was unbroken, did not curve too much, and was clean and uniformly lit. Schmidt noted that handling even simple exceptions, such as brightness changes caused by shadows, would require several times as much computation to consider the wider alternatives. Detecting and responding to obstacles, road signs and other hazards promised to be much more complicated.

Shakey's vision programs, as most others of the time, reduced images to a short list of geometric edges before doing anything else. The approach was quite inappropriate for outdoor scenes containing few simple edges, but many complicated shapes and color patterns. A major exception to the blocks-world approach was a project begun at SAIL with impetus from Joshua Lederberg, then a Stanford geneticist with a Nobel prize, and John McCarthy's enthusiastic support. It was to look for changes on Mars that might indicate growing life. Using digital images from Mars-imaging spacecraft Mariners 4, 6, 7, and 9, the project worked to register, in geometry and color, views of the same regions taken at different times so that any differences could be detected. Since the spacecraft locations were known only approximately, the image registration process was to be guided by surface features themselves. A graduate student, Lynn Quam, and others developed a collection of statistical, intensity-based comparison, search, and transformation methods that did the job (alas, no unambiguous life signs were spotted). Since they dealt with complex natural scenes, the Mars methods also seemed appropriate for interpreting imagery from an outdoor vehicle. The NASA Mars rover program was then in the ascendant, so there was a double bond between the Cart concept and the Mars group. I arrived in late 1971, an enthusiast for space and robots, and soon adopted the Cart from Bruce Baumgart, another graduate student who had been maintaining it for possible use in his own research in computer graphics and vision. The Cart had little research reputation, but discussions with Quam produced a plan where I would provide a working vehicle (a nontrivial project given the shoestring construction) and PDP-10-resident motor control software. Quam and company would adapt their image methods for visual navigation. By 1973 I was having a good time building and test-driving new robot hardware and software, when, in a remote-control misstep, I crashed the poor Cart off a small loading ramp. Months of low budget repairs left its TV transmitter still broken and led me to beg McCarthy to invest several thousand dollars in a replacement. He agreed, but insisted that I first demonstrate competence in writing programs for computer vision.

Real-time performance was not an issue in interpreting Mars images. The missions were several years apart, and, until the Mariner 9 orbiter, each produced only a few dozen images. The Mars group could afford to run search programs for hours at a time to find precise and dense matches over large image areas. A Cart-driving program might forgo this precision and coverage in exchange for speed. It seemed that many tasks could be accomplished with just two basic image operations—one to pick out a good collection of *features* (distinctive local regions peppered across a scene) and another to find them in different views of the same area. Three-dimensional locations could then be determined by triangulation, obstacles detected, and the motion of the robot deduced. I set about to find fast implementations of these ideas. Doing the bulk of their work in spatially compressed images—where squares of 4, 16, 64, and more pixels were averaged into one—and cleverly coded in assembly language, my operators were able to pick out a few dozen good features in one image and reacquire them in another using about ten seconds of computer time. In 1975 I built a program around them that controlled the Cart's heading by tracking horizon features on the roads around SAIL. The program would repeatedly digitize a frame and, in fifteen seconds, determine the horizontal displacement of features on the (usually tree-lined) boundary between ground and sky between frames, calculate a steering correction, and drive the robot up to ten meters. It did its unambitious task quite well, and was fun to watch, but it was intended as mere practice for the main event: a much more ambitious program that would drive the Cart through an obstacle field by visually tracking its surroundings in three dimensions—to build a map, identify obstacles, plan safe routes, and—most difficult—deduce and correct the robot's motion from the apparent motion of those surroundings. I decided to approach this task in full three dimensions from the start, hoping eventually to run the robot on the rolling adobe terrain outside the lab.

The Cart carried a single camera, so it was natural to use driving motion to provide multiple viewpoints for triangulating distances. Complicating the matter, the Cart's motors were very imprecise, so its moves would have to be deduced simultaneously with the position of tracked objects. The Mars team had a similar problem, and another student, Donald Gennery, had already written a "camera solver" for it. I struggled with this approach through early 1977. The program would take a picture and choose up to a hundred features. It would then drive the robot forward about a meter, stop, take another picture, and search for the same features in the second image. Then it would invoke the camera solver to find the robot movement and the three-dimensional locations of the features that explained their apparent motion from one image to the other. Despite much fine-tuning, the program's error rate never dropped below about one wrong motion solution in four, meaning the robot could move about four meters before becoming confused about its position—discouraging. The camera solver repeatedly tweaked an estimate of the robot's motion to make the features line up as well as possible, and threw away those that seemed too far off. It worked well for high-quality spacecraft images of an almost two-dimensional surface with few matching errors, good initial camera position estimates, and ideal sideways motion between images.

My data, with poor position estimates from noisy TV images of a nearby scene, with plenty of perspective distortion from frame to frame, was something else. Ten to twenty percent of the feature matches were wrong, often because an area chosen in a first image had, in a second image, been eclipsed or changed in appearance by point-of-view or lighting effects or camera noise. Position accuracy in my low-resolution images was modest, compounding the serious limitation that forward motion stereo is mathematically insoluble for points near the camera axis. The combination of many outright incorrect matches and large uncertainties in the correct ones made finding the robot motion a chancy proposition. It was necessary to track about one hundred features to succeed even three steps in four, consuming several minutes of computer time. Months of fiddling with the program's mathematics and assumptions made little difference. Eventually, I chose to add some robot hardware to reduce the uncertainties.

Multiple cameras or a repositionable camera on the robot would permit true stereoscopy, from precisely located relative views, removing the biggest source of uncertainty. Mismatched features between stops might then be pruned, before solving for robot motion, by exploiting the constraint that the mutual three-dimensional distances between pairs of features should remain unchanged by a move. Vic Scheinman, an engineer and graduate student whose main interest was robot arms, but who often graciously lent his mechanical expertise to the Cart project, found for me, in his basement, a mechanism able to slide the camera about 60 centimeters from side to side. Motorized, this provided a fine stereo baseline. Errors were further reduced by taking pictures at nine places along the track, exploiting the nine-way redundancy.

The final result, first sufficiently debugged in October 1979, was a program that would track about thirty small image features at a time to drive the robot through indoor clutter, avoiding obstacles and accumulating a sparse map of the scenery. Its computer (now a 1 MIPS KL-10) worked for ten minutes to prepare for every meter-long move. In five hours it would arrive at a requested destination at the opposite end of a 30-meter room, succeeding in about three traverses in four. In outdoor tests the Cart managed to travel only about 15 meters before becoming confused. Harsh contrast between sunlight and shadow overwhelmed the old-style TV camera tube and greatly degraded its vision.

When the navigation failed, it was usually because of the process intended to prune errors made in tracking features from one stop to the next. The nine-eyed stereoscopy located features quite reliably in three dimensions at each stop, relative to the robot's position and orientation. Though the reference frame changed after a move, the mutual 3D distance between pairs of features should not. The pruning process aggressively strained out those features that violated this "rigidity" criterion, up to half of the one hundred raw matches. But sometimes, by chance, about once in one hundred moves, some of the mismatched points would happen to mutually support one another more strongly than the correct ones, and the pruning would retain them, and reject the good matches! The robot would then mis-estimate its position and heading, mess up its accumulating map, and run into trouble.

Gridwork

In 1980, Ph.D. done, I moved to Carnegie Mellon University and set up a small "mobile robot laboratory," where we built and worked with new robots. My first two students, Larry Matthies and Chuck Thorpe, streamlined the Cart visual navigation program by using fewer images and exploiting constraints, like the robot's flat indoor floor, to simplify and speed up the program tenfold. They increased its navigational accuracy by modeling geometric uncertainties more precisely. The changes hardly altered the once in one hundred navigational failure problem. Apparently, given the matching error rate, one hundred features gave bad luck too much of an opening. The chance that random errors could overpower good data was significant in that small a sample.

In 1984 our laboratory accepted a research contract from a small new company, Denning Mobile Robotics, challenging us to navigate robots using range measurements from an obstacle-detecting belt of twenty-four sonar units. Computer vision then was much too expensive for an affordable machine, but the sonar devices, developed for autofocusing Polaroid cameras, cost only a few dollars.

Each sonar measurement is the echo time of a sound pulse in a 30° wide beam, and although the distance is quite accurate, there is no indication of where, laterally in the huge 30° field of view, the echo originates. The Cart program's techniques, dependent on pinpointing distinctive features in the scene, could not be used, even if they were reliable or fast enough. Another student, Alberto Elfes, and I devised a completely different approach that, instead of trying to catalog the location of objects, accumulates the "objectness" of locations. Whereas the identity or even existence of particular objects is always in question, locations around the robot can safely be assumed to exist, and can serve as permanent "buckets" to steadily accumulate even a light rain of evidence about their contents.

The area around the robot is divided into a grid. A program maintains a number for each cell of the grid representing the evidence accumulated thus far that the corresponding cell contains something or, contrarily, is empty. With each new sonar ping, cells representing its sweep in space are altered. Cells at the echo distance gain evidence of occupancy (because somewhere at that distance is the cause of the echo), while those in the interior of the beam lose occupancy evidence (because anything in the interior volume would have caused a shorter echo). The amount of evidence adjustment varies across the beam's volume because the sound intensity, and so the sensing reliability, declines gradually away from the center of the beam, and with distance.

Given the problems with the Cart approach, we were very surprised in 1985 when our first program using the evidence grid method enabled a robot to build maps of its surroundings and cross our cluttered lab with almost perfect reliability. But was it practical? A reasonably fine three-dimensional grid divides a room into several million cells, and each sonar measurement affects tens of thousands of those, demanding more computer memory and speed than could reasonably be installed on a robot in 1985. By using a very coarse grid, with cells 30 centimeters on a side, in only two dimensions, like a map, we produced an efficient program able to handle ten sonar readings per second on the then–1/2-MIPS Denning robot. A key navigation step—aligning two maps of the same area—took 3 seconds. Close, but not good enough for a fast-moving robot. A future robot model with more powerful computers could probably handle it, but the company was stretched too thin to immediately pursue that option.

In the meantime, we continued exploring this promising path with funding from the Office of Naval Research. We applied the grid approach to distances extracted from stereo vision, then combined stereo and sonar data in a single map. We put its mathematics on a sounder foundation of probability theory. We acquired more powerful computers and did more experiments. One of these revealed a weakness: in a narrow, smooth-walled hallway—unlike our very cluttered lab—sonar pulses ricocheted around, like light in a hall of mirrors, most readings were misleading, and the resulting maps were worthless. The evidence patterns the program added to its map for each reading—which we had constructed by hand from the specifications of the sonar units—were not indicative of what the sonars were really saying about the hallway.

In 1990 we developed a "learning" approach to find better evidence patterns. The patterns were encoded as mathematical formulas with a dozen parameters—"knobs" controlling their shape. We carefully measured the hallway by hand and constructed a near-perfect *ideal* map of it. We ran a robot down the hallway in a precisely known way, and collected sonar data at regular intervals. Then we wrote a program to repeatedly process the collected data to simulate a map-building robot traveling down the hall. After each simulated run, the program compared the resulting map with the ideal. Then it tweaked the knobs on the evidence formula and simulated another run. If the map using the new evidence patterns was more like the ideal than before, the program moved the knobs further in the same direction; otherwise it adjusted them the other way. During the course of hundreds of thousands of simulated robot excursions, consuming days on machines that by then had reached 10 MIPS, the program gradually "tuned in" an evidence pattern that produced an excellent representation of the hallway, and was suitable for other smooth-walled surroundings.

Heightened Sense

On 10 MIPS computers, two-dimensional evidence grids can be built and used quickly enough for normal indoor speeds. In the 1990s a growing number of research groups worldwide are fielding robots that trundle around office settings, guided by 2D grids and other kinds of maps of similar complexity. Though impressive as short-term demonstrations, these machines invariably run into problems, like colliding, or becoming lost or trapped, perhaps several times a day. Although they are much better than sparse rep-

1990—2D evidence grid of a difficult hallway
A robot with a belt of 24 Polaroid sonar units obtained 624 range measurements along an 8 meter hallway. More than half the ranges were too long or missing because of deflections by the mirrorlike walls. A good reconstruction, in a 64 by 32 cell grid, was nevertheless obtained because the evidence patterns representing sonar ranges had been well adjusted for this kind of environment by a learning process. The traversed hallway runs left-right. The start of an adjoining hallway, running upwards, can be seen on the right.

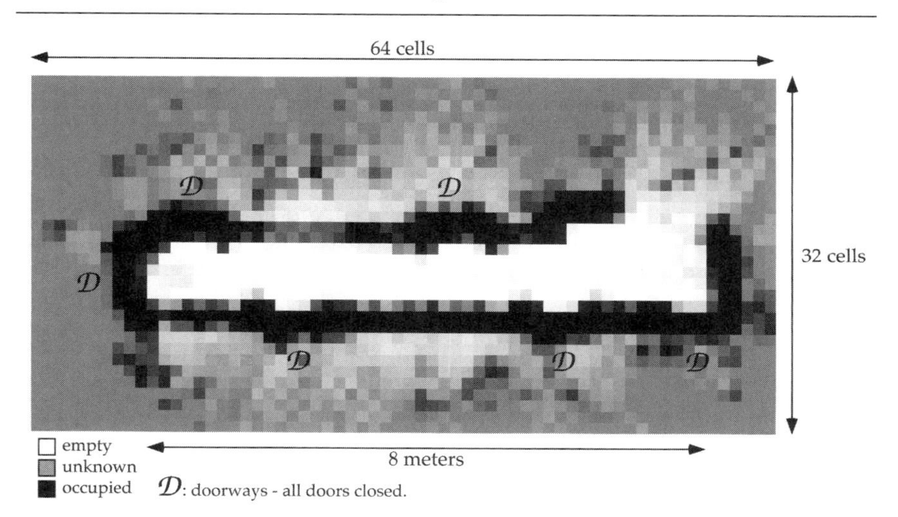

resentations like the Cart's few dozen features, maps with a few thousand cells still have a significant probability of being fooled by unlucky combinations of sensing errors.

The probability of misapprehension drops as the number of independent things known about the world grows. A most attractive way to increase the information in evidence grids is to do them in full three dimensions. In two dimensions doorknobs, desktops, and other things that vary with height are sensed inconsistently, and only a blurry kind of map is possible, so there is little advantage in reducing the cell dimensions below about 10 centimeters on a side. In 3D the world is consistent and there need be no blurring, so the grid resolution can be higher and its cell contents more certain. In 2D a chair is a fuzzy blob a few cells across, indistinguishable from other objects of similar size. In 3D a chair can have legs, seat, and back, and be recognizable by shape. A 3D map would enable a program to plan a complex path that weaves not only around but over and under obstructions. The chance of navigational misperception could become negligible.

The price is high. A 2D grid, which we can manage with 10 MIPS, has a few thousand cells. In 3D it is tempting to greatly reduce the cell size to a few centimeters or less. Quadrupling the resolution of a 2D grid increases its number of cells by a factor of 16. In 3D, the effect is compounded by the third dimension, which itself may be 100 cells high, resulting in several thousand times as many cells to store and process in 3D as in 2D. It seemed 3D needed computers with over 1,000 MIPS.

In the early 1990s, 1,000 MIPS could be found only in supercomputers. In 1992 I spent a sabbatical year in Boston as a guest of supercomputer maker Thinking Machines Corporation. My intent was to write a program to project evidence rays into 3D grids using the new CM-5 machine, composed of several hundred 20 MIPS computers working in tight concert. My desired programming environment was not ready, so I used a regular computer workstation to develop the programs that would eventually run in many copies on the CM-5.

That preamble grew into an eight-month project. Large grids made possible economies of scale. Just about every sensor could be well represented by evidence rays symmetric around an axis of propagation, and thus could be represented by 2D radial slices of cylinders, rather than 3D boxes. For each new measurement, the transformation that rotated these slices into cylinders skewering 3D grid maps was very similar from layer to layer in the grid. It could be worked out for one slice in a very efficient and compact form and repeatedly reused. In each slice, the cells could be sorted by radius from the propagation axis and then used only up to the maximum radius of actual data at the slice. Only the typical cone of evidence rather than the whole cylinder needed to be filled. Perfecting these and other ideas produced a program about 40 times as efficient as anticipated. A surprise speedup came from an advance in the compiler, which translates programs into machine code. Setting the optimization of "Gnu C" to level 3 sped up the code about 2.5 times. All together, the final program ran 100 times faster than expected. Further, my 1992 workstation computed at 25 MIPS, giving yet another 2.5 factor. Equally important, the workstation had enough memory, about 16 megabytes, to store an entire 3D grid and its supporting structures. After eight months of preparing to use a supercomputer, I no longer needed one!

At 25 MIPS, in a cubic grid 128 cells on a side, the program added 200 broad sonar-like measurements per second, or 4,000 thin rays, as might represent laser or stereoscopic range values. It was fast enough for developmental experiments, if not quite for working robots.

The 1992 sabbatical had left me with an unexpectedly practical core of a 3D grid program, but not an entire robot perception system. Groundwork and distractions stretched the next step over several years. Sonar, providing only hundreds of fuzzy measurements per minute, was not a promising sensor for a high-resolution 3D grid with millions of evidence-hungry cells. Scanning laser rangefinders, used in a simpler way by some other robot projects, could provide the requisite data, but were large, expensive, balky, and power hungry. I imagined our methods showing up in future small inexpensive mass-produced robots and looked for something more practical. Television cameras had shrunk to finger size, so stereoscopic vision again looked attractive.

Wide-angle lenses give nice coverage, but they introduce a pronounced "fish-eye" distortion. By 1995 I had written a program to correct such camera imperfections. The program works by viewing a precise calibration pattern of several hundred black spots through a camera, from which it derives a "rectification" function that straightens out the raw image into one with ideal geometry. Still, the work was dragging. Since my last sabbatical had been very productive, I accepted the opportunity to do another in 1996, this time at Daimler-Benz research in Berlin, Germany.

My workstation in Berlin had 100 MIPS and 64 megabytes. I wrote a stereoscopic front end for 3D grids, using methods resembling those of the Cart program twenty years earlier, but improved in many subtle ways. Input came from two cameras side-by-side on a tripod moved by hand (there was a robot in the next room, but not quite ready to host my programs). Instead of extracting a few dozen features from each stereo set, the new program extracted about 2,500. Every found feature was translated into at least two evidence rays, one from each camera, to the triangulated location of the feature. A grid 256 by 256 by 64 cells represented a volume 6 meters wide by 6 meters deep by 2 meters high.

With economies of scale in the stereoscopic ranging as in the grid updating, the program was able to process a pair of images and add the consequent 5,000 evidence rays to the grid in about five seconds.

The following image contrasts the new result with one from the Cart. Each map was derived from about 40 separate camera views of the scene. The Cart map took about 40 minutes of computation at 1 MIPS with 1/2 megabyte of memory, while the grid required 80 seconds at 100 MIPS and 20 megabytes. The Cart map marks the location of about 50 features, while the grid indicates about 100,000 cells are occupied (also that 1.5 million are empty and that 2.5 million remain unknown). The Cart map is barely usable for navigation. The grid map promises to support not only highly reliable navigation, but the recognition of objects by 3D shape.

The 3D grid results are very encouraging, everything I had hoped for, but open many avenues for improvement. An interesting one concerns learning the evidence patterns to best represent individual measurements, an approach that greatly improved 2D sonar maps. The patterns are now devised from the stereoscopic geometry and many guesses. But how to guide the learning process? In 2D the learning evaluated computed grid maps against a hand-constructed ideal map. The thousandfold greater number of cells makes constructing a 3D ideal impractical. But although we don't have an ideal 3D model, we have excellent pictures of the 3D scene in the original stereoscopic images! The occupied cells of a computed grid can be viewed, as in the illustration, from perspectives corresponding to the camera images, and compared to them. The better the match, the better the grid, if only the grid cells had colors corresponding to the image. In fact, the grid can be colored by sometimes projecting the image colors onto the cells, instead of matching them. Not only will this colorizing and matching approach allow the grid program to tune itself up at any time, but the color will enhance object recognition and other operations.

The work goes on. A new graduate student, Martin Martin, is

1979 and 1997—3D maps from stereoscopic images

In each example about forty images similar to the ones in the upper half of the diagrams were stereoscopically processed to make the maps below them. The map on the left shows the Cart's position near the beginning of a run, its camera field of view, its planned path, and a kind of perspective view of features that passed all consistency checks. Each feature is a black dot linked to the ground by a diagonal line. The diagonal's length is the feature's height; where it meets the ground is its forward and lateral position. The features are also marked by dots overlaid on the camera image. To aid interpretation, about a dozen clusters of features were hand-labeled with the identity of the object on which they were found. The map on the right is a perspective view of the occupied cells in a 256 by 256 by 64 grid representing a 6 meter square by 2 meter high bite of the office above. To aid visualization, the cells in about a dozen box-shaped volumes selected by hand were "spotlighted" with distinctive tints.

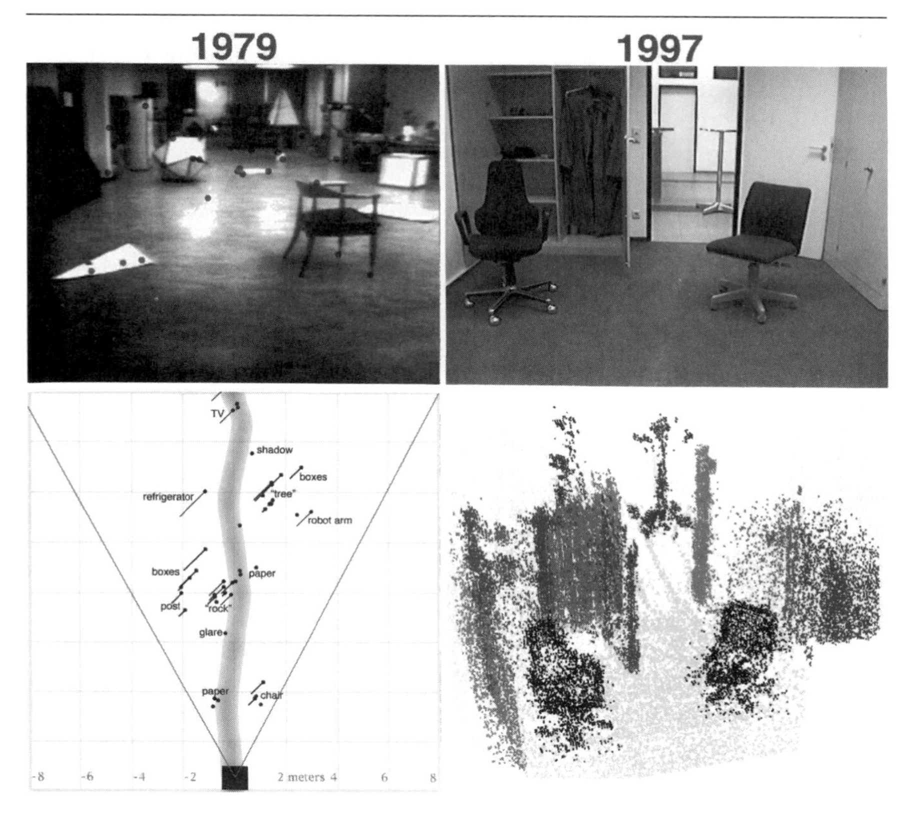

using 3D grids in a robot whose computer will soon be upgraded to 500 MIPS. We hope to demonstrate navigation for practical near-term possibilities like Chapter 4's free-roaming robot vacuum cleaners.

But what about John McCarthy's self-driving cars? Our lab's slow little robots, even when perfected, would be roadkill if they ventured onto a street. But less timid machines have emerged to meet the challenge.

Fast Cars

Minicomputers small enough to fit into cars appeared in the 1970s, and a few actual computer-driven cars and trucks appeared within a decade. As with small robots, their performance remained discouraging until the 1990s.

In 1977 Japan's Mechanical Engineering Laboratory built a stereoscopically guided autonomous automobile that could follow well-defined roads for distances of about 50 meters at up to 30 km/h. The secret was highly specialized hardware filling a rack on the passenger side of a small car, using input from two small television cameras mounted sideways one above the other on the car's front grill. The cameras' video signals were electronically filtered to detect brightness changes that were then registered as digital pulses. The pulse streams from the two cameras were matched at various lateral displacements by simple digital circuitry: displacements with many matching pulses indicated objects at certain distances. When properly adjusted for lighting and contrast, this specialized circuit, doing the equivalent of about 50 MIPS of computing, gave the distance of about eight major visual discontinuities, such as road embankments and obstacles, thirty times per second. The distances were sampled ten times a second by a 1/4-MIPS minicomputer programmed to keep the vehicle on the road and veer around obstacles. The system performed impressively when the edge detectors were carefully adjusted, and the road and obstacle boundaries were contrasty enough—the test roads were usually edged by white lines. At other times, the vehicle was unpredictable and dangerous. The problems were fundamental to the approach, and the project was ended in 1981.

In 1984, as part of a larger program called the Strategic Computing Initiative, the Department of Defense Advanced Research Projects Agency (DARPA),[8] initiated an overambitious program called "Autonomous Land Vehicles" (ALV). It promised stealthy robot crawlers to do battlefield reconnaissance, sabotage, and perhaps combat. The prior decade of computer vision work had convinced the managers at DARPA that stereo vision was too hard a problem for their timeframe, but they guessed that the rest of the perception and navigation problem was tractable. Before being abandoned in 1989, the project had financed a half-dozen small experimental vision-guided vehicles and two big ones. A rough-terrain vehicle as big as a bus at Martin-Marietta Corporation in Denver was equipped with about 50 MIPS of specialized computer power, color television cameras, and a scanning laser rangefinder that provided, twice a second, a 128-by-256 array of distance measurements, doing by optics and electronics what was impractical by computation. A similar machine, a large Chevy van, heavily modified, was developed at Carnegie Mellon University. By the end of the project, Martin-Marietta's ALV and Carnegie Mellon's Navlab were both driving down dirt roads at speeds up to 20 km/h, but usually much slower, tracking road boundaries in color video images, and stopping for obstacles detected by a minimal processing of the laser range data. Both were able to do this as long as the road boundaries were relatively well defined and without major discontinuities. The simple road identification techniques were often fooled, and both were unlikely to stay on a road for a whole kilometer.

A project at the Bundeswehr University in Munich funded partly by German automobile and electronics companies began in 1984, and by 1989 had produced a van that sometimes drove autonomously on the autobahn at up to 100 km/h, guided by a TV image processed by an onboard array of a dozen special-purpose computers, each—at 10 MIPS—able to track a single image patch at 13 frames per second. The features were selected by a human at the beginning of an autonomous run, typically one at the left edge of the highway or the lane, another at the right edge, and others chosen on license plates or other distinctive marks on traffic ahead and to the sides of the van. Using motion-prediction techniques,

Carnegie Mellon's self-driving Navlabs
Numbers 1 to 5, left to right, a sequence that stretches from 1986 to 1996, 1 meter/sec to 150 km/h, 10 meter to 5000 km, 1 MIPS to 100 MIPS, refrigerator-sized to laptop—and half the fun was getting there.

the image processors maintained their visual locks for many minutes. Their output went to a minicomputer programmed for motion control, to keep the vehicle in the lane, at a safe distance from the other traffic. In 1989 the van ran the entire length of an empty 20 km autobahn spur.

The above systems critically depended on an alert human supervisor with a manual override button—these simple-minded and dangerous machines were very easily confused by such common driving events as shadows, road stains, lampposts, stopped cars, or sudden curves.

Carnegie Mellon's Navlab project outlived the ALV program. Managed by mobile robot lab alumnus Chuck Thorpe, who had earned his Ph.D. in 1984, and funded by automotive, farming, and mining industry as well as government contracts, it expanded into a range of applications on roads and fields and mines. The road work was particularly instructive.

In 1984 Navlab's predecessor, the Terragator, a desk-sized outdoor robot, radio-controlled by a housebound 1-MIPS computer, crept along jogging trails at about a meter per second. Occasionally it mistook a tree trunk for the road and tried to climb it! The Navlab, running no faster, replaced it in 1986. It was a big blue truck crammed front to back with racks containing 1 MIPS of computing, a huge air conditioner, and a heavy generator. The original drive train was replaced by hydraulics for control at very slow speed. A succession of handcrafted programs to pick out the boundaries of the road ahead in the color camera view elicited a few general tricks (e.g., subtract the green from the blue channel: the road is bluish from reflected sky, the roadside greenish from vegetation; subtraction enhances the distinction and eliminates shadows) and slowly improving performance. In 1988 the van negotiated a network of empty city streets at a few kilometers an hour. By 1990, with about 10 MIPS of improved computers, it sometimes reached the 40 km/h top speed of the hydraulics. Still, almost every time a new road type or condition was encountered, the programs needed tinkering.

In 1990 Navlab was replaced by Navlab 2, a humvee that had been a military ambulance, containing three 20-MIPS workstations. Thorpe's student Dean Pomerleau introduced a new approach. Instead of handcrafting the road finders, Pomerleau's "ALVINN" program trained a simulated neural net, with about 5,000 adjustable interconnections, to imitate a human driver. The net's input was a low-resolution image of the road, preprocessed with the blue-from-green subtraction trick. Its output determined the steering wheel position. In early versions, the net required hours of training, provided by simulator scenes or long videotapes of human road trips. Training tricks gradually reduced the time. Each original image was expanded into dozens. Some simulated the vehicle being further left or right in the lane, with corresponding adjustments to the steering, providing experience outside the boundaries of normal operation. Others added random clutter in the off-road locations, training the net to ignore the contents of those regions. Ultimately, the time to learn new roads was reduced to about five minutes. For long trips, the system was provided with the neural interconnection weights for many road types. It tested each against the current road, using consistency of the steering output as a measure of suitability. In 1991 ALVINN traversed 30 km of busy highway at 70 km/h, and Pomerleau received his Ph.D.

In 1995 another Navlab, another student, and another idea topped this performance significantly. Navlab 5 (3 and 4 are unimportant) is a Plymouth minivan. Instead of being crammed full of equipment, it is controlled by a 50-MIPS laptop computer. The camera, rather than being ostentatiously mounted on the cab roof, peeks out the windshield over the rearview mirror. The student is Pomerleau's, Todd Jochem. The idea was to replace ALVINN's general learning, reined in by a large quantity of contrived training, by a specialized system in which the known constraints were built in, and only the changeable parts were learned. "RALPH" begins with the same 32-by-32 pixel low-resolution road picture as ALVINN. In it, the lane ahead of the vehicle, tapering into the distance, appears as a wedge. RALPH distorts the image to stretch this wedge into a straight vertical ribbon, then averages the ribbon down to a single 32-pixel vector, representing an average cross section of the lane. Of course, this is true only if the lane lies straight ahead. If it angles right or left, or curves, the stretching and averaging will blur together different parts. So the program also tries transformations representing a range of angles and curves. For each hypothesis, the program estimates the blur in the resulting vector from the brightness changes from one cell to the next: blurring softens such changes. The sharpest vector is kept; the others are discarded. To learn a road type, its 32-number vector is instantly recorded (compare this to ALVINN's 5,000 numbers learned over 5 minutes). In automatic driving, the current vector slides across a memorized vector, and where it best matches indicates where the vehicle is in the lane. On a new road, the current vector can be checked against a library of thousands to find the best match. RALPH can even train itself for new road types by comparing a vector made from only the lower half of its image to the vector from the upper half. If they match, the road is the same up ahead as locally. If not, the upper half is a new road type, which can be added to the library. The program has driven successfully in the dark, in heavy snow, and in driving rain. When visibility is near zero, it can lock on to weak or transient features like pavement stains or water ripples from preceding cars! In the summer of 1995 RALPH drove Navlab 5 from Washington, D.C., to San Diego, California, in control 98.2% of the time, at an average speed of over 100 km/h. Jochem got his Ph.D. in 1996.

Another generation of students is now building this work into the "highway of the future" and "driver's assistant," among other applications.

Night Crawlers

Only a handful of computer-controlled robots existed while computers were very expensive, but small, cheap microprocessor "brains" sparked a robot euphoria in the 1980s. It seemed that robotics might repeat the amateur computer boom of the late 1970s. Thousands of hobbyists, small companies, industrial and government labs, high schools, university undergraduates, and graduates worldwide built or purchased small programmable vehicles. Heathkit sold 25,000 "Hero" robots costing about $2,000 each, and a half dozen other companies sold thousands of their hobby robots. The low-budget majority of these machines relied on contact, sonar, or infrared proximity detectors and fractional MIPS processors to execute preprogrammed sequences with simple reactions to obstacles and a few other conditions. A few dozen larger and more expensive robots carried TV cameras or laser rangefinders and often several 1-MIPS processors. A very few had onboard manipulators.

The repertoire of the small robots was soon exhausted, and the mobile robot boomlet ended before 1990. A few dozen more serious projects were left in its wake. Some used sonar or optical range sensors to build two-dimensional line maps of their surroundings. These could work in real time on robots moving slowly in simple office environments but were overwhelmed in cluttered spaces or where sensors had a high error rate. Some computer-vision research groups acquired camera-equipped robots and used stereo vision, range stripe, and shading-based scene interpretation approaches—methods that consumed minutes or hours of computer time for each glimpse of the world. Often they concluded that a mobile robot is an expensive and troublesome way to obtain a few pictures.

Only in the 1990s, as their onboard computers exceeded 10

MIPS, did complex mobile robots really begin to show some promise. Today there are dozens of wastebasket-sized machines wandering the halls of universities and other research institutions all over the world. A smaller number of other robots fly, swim, crawl on legs, and of course drive in the great outdoors, their visual senses augmented by satellite and inertial navigation instruments. They're not quite ready for general use, but are good enough for occasional competitions, and entertaining enough to be featured in dozens of television broadcasts and magazine articles. And each year their average performance improves.

The frustrating state of "model-based" (or "map-based") robotics in the late 1980s generated one particularly radical reaction. Rod Brooks at MIT, with great energy and showmanship, declared that model-based approaches were unworkable, and demonstrated how to get complex behaviors from robots *without* models. Brooks's machines were controlled by cleverly contrived layers of reflexes, or *behaviors*, like Grey Walter's tortoises or the Hopkins Beast, only more elaborate. One behavior might cause the robots' motors to steer away from obstacles sensed by proximity, while another would use the same sensors to keep the robot traveling along a wall. Some behaviors responded to others, or overrode them. Each behavior was a simple program sharing a modest processor; typically the whole robot might contain 1 MIPS hosting a few dozen behaviors. Insects seem to be controlled this way, each behavior produced by ganglia of a few hundred neurons. Brooks's robots acted very much like insects, a similarity emphasized when his lab began to make miniature robots that walked on six legs. They were more lifelike than most model-based robots and less predictable. After the novelty wore off, however, it had to be admitted that behavior-based robots did not accomplish complex goals any more reliably than machines with more integrated controllers. Real insects illustrate the problem. The vast majority fail to complete their life cycles, often doomed, like moths trapped by a streetlight, by severe cognitive limitations. Only astronomical egg production ensures that enough offspring survive, by chance.

In the 1990s it is model-based machines that are navigating offices and crossing the country. They have just a few reflex behaviors, mostly related to avoiding obstacles and other immediate dangers. Perhaps in reaction, Brooks has raised his sights to "Cog," a humanoid robot controlled by large numbers of learned reflexes that, to date, allow it to learn to visually track and grasp an object. It has direct analogies with our own nervous system, but I think imitates it at too low a level. The Cog approach, given enough time and effort, probably could retrace our own evolution and development and lead to a fully intelligent machine. I think there's a much faster route, though, one that imitates at a much higher level of abstraction, pragmatically exploiting every engineering and computer science trick that helps. It's outlined in Chapter 4.

Nevertheless, of necessity, most practical automatic machines are behavior-based. Although designed to be predictable rather than lifelike, the cunningly wired racks of electromechanical relays that have controlled automatic elevators, streetlights, and hosts of industrial machines for fifty years are like very simple insect ganglia. Since the 1980s, many relay boxes have been replaced by microprocessors, at first programmed simply to mimic the relays. The complexity has slowly grown, but most controllers still react to stimuli like touch, light, temperature, and time in simple ways, carefully chosen to be effective.

Will Work for Electricity

The robot giddiness of the early 1980s generated not only a horde of hobbyists, but also a handful of little companies who promised to deliver human-sized microprocessor-controlled mobile robots for institutional use. These were to be smaller, cheaper, smarter, and more broadly useful than the large Automatic Guided Vehicles (AGVs) that had ferried materials from place to place in some factories since the 1950s. The older machines, costing about $100,000, reflexively followed signal-emitting wires buried under concrete factory floors, and started and stopped in response to switch contacts and simple timers. In the late 1970s some AGVs were redesigned with computer controls and the means to navigate by painted lines or other contrasting floor marks.

The new mobile robots aspired to be as advanced as the new generation of computer-controlled small electric arms. Practical, flexibly programmed sensor-guided electric manipulators were first designed by Vic Scheinman, a graduate student in mechanical engineering at the Stanford AI Lab "hand-eye" project in the early 1970s. They were commercialized and used for precision light assembly in the automobile and electronics industries by the end of the 1970s. Technically impressive machines though they were, versatile, responsive, fast, and precise, they floundered commercially. For most jobs, robot state-of-the-art is still inferior to human labor in cost and performance, and advanced robotics has been a small and unprofitable business. There are a few exceptions in Japan, where manufacturing companies, anticipating a labor shortage in an aging population, have purchased many robots to familiarize themselves with the technology for the future.

Mobile robots proved just as difficult to make and market as robot arms. Several dozen small robot-making companies, and divisions in large companies, appeared during the 1980s, only to disappear a few years later. Today a few are still hanging on, hoping the climate will soon improve. Transitions Research Company (TRC) in Connecticut, and Cybermotion in Virginia hope to make millions of robots someday, but for over a decade have survived by raising modest sums from investors and selling about a dozen $25,000 robots a year, mostly to research groups. Each company has also sold about one hundred robots for actual end-use. TRC's robots deliver meals and linens in hospitals, Cybermotion robot security guards patrol warehouses. Their navigation, conceived in the 1980s on 1/2-MIPS computers, orchestrates simple reflexes, for instance, to avoid obstacles, follow along walls, or head toward beacons, into a large plan that carries them around their workspace in a desired way. The biggest problem for these end uses was making the robots ultrareliable. The successfully installed robots have run their assigned routes for many years. In contrast, a TRC cart tested at the Stanford Hospital in 1992 lost its job after a month of continuous service, when it failed to respond to a transmitter marking a "no robot zone" and worked its way down a stairwell! With existing techniques, such reliability is tedious and expensive to achieve. Programs for negotiating each segment of hallway, each turn, each door must be carefully prepared, adjusted, and tested. Special navigational markers may have to be affixed at key locations.

The need to call a specialist every time the robot must be

rerouted dismays potential customers. Still, there is interest. Robots that could both learn routes easily and execute them reliably, even under adverse conditions, would find many uses. Such machines have proven harder to create than expected, but they are almost here.

3

Power and Presence

Like a desert in a cloudburst, the entire field of robot intelligence is sprouting after thirty arid years. What irrigates the seeds, how fast will they grow, and into what?

Computational power is to a mind what a locomotive engine is to a train. The train can't move if the engine is too small. But engine power is effective only if properly coupled to the load. Locomotive engineers of the eighteenth century learned the relationship between speed, pulling power, engine size, and transmission ratios by trial and error, no doubt overturning many horsecart-derived intuitions.

Two centuries later, robotics is learning analogous lessons. The first electronic computers were as big and expensive as locomotives, and pulled arithmetical loads like armies of humans. We can forgive researchers of the 1960s for feeling that these behemoths could pull other kinds of mental cargo equally powerfully, if hitched via the right programs. Alas, the thirty intervening years have shown that intuition to be wildly miscalibrated.

Humans chug furiously to accomplish the unnatural act of calculating. Images, sounds, and feelings cascade through our brains for minutes to produce such nuggets as "35 times 237 makes 8295." Similarly, a tractor roars, diesel fumes belching, to creep along a road, with gears configured for plowing, not speeding. A computer besting a human at calculation is like a bicycle speeding past the tractor. The bicycle has the advantage on the road, but oh what an exercise in frustration trying to find the right rigging to enable the bicycle to pull a plow! Only a monster bicycle, built for 1,000, could match the tractor. Similarly, only a monstrously powerful computer could hope to match humans in natural mental activities like perceiving, moving, and socializing. Let's attempt to quantify "monstrous."

Brains, Eyes, and Machines

Computers have far to go to match human strengths, and our estimates will depend on analogy and extrapolation. Fortunately, these are grounded in the first bit of the journey, now behind us. Thirty years of computer vision reveals that 1 MIPS can extract simple features from real-time imagery—tracking a white line or a white spot on a mottled background. 10 MIPS can follow complex gray-scale patches—as smart bombs, cruise missiles, and early self-driving vans attest. 100 MIPS can follow moderately unpredictable features like roads—as recent long Navlab trips demonstrate. 1,000 MIPS will be adequate for coarse-grained three-dimensional spatial awareness—illustrated by several mid-resolution stereoscopic vision programs, including my own. 10,000 MIPS can find three-dimensional objects in clutter—suggested by several "bin-picking" and high-resolution stereo vision demonstrations that accomplish the task in an hour or so at 10 MIPS. The data fades there, awaiting computers with more speed and more memory.

There are considerations other than sheer scale. At 1 MIPS the best results come from finely handcrafted programs that distill sensor data with utmost efficiency. 100-MIPS processes weigh their inputs against a wide range of hypotheses, with many parameters, that learning programs adjust better than the overburdened programmers. Learning of all sorts will be increasingly important as computer power and robot programs grow. This effect is evident in related areas. At the close of the 1980s, as widely available computers reached 10 MIPS, good optical character reading (OCR) programs, able to read most printed and typewritten text, began to appear. They used hand-constructed "feature detectors" for parts of letter shapes and did very little learning. As computer power passed 100 MIPS, trainable OCR programs appeared that could learn unusual typestyles from examples, and the latest and best programs learn their entire data sets. Handwriting recognizers, used by the Post Office to sort mail, and in computers, notably Apple's Newton handheld, followed a similar path. Speech recognition also fits the model. Under the direction of Raj Reddy, who began his research at Stanford in the 1960s, Carnegie Mellon has led in computer transcription of continuous spoken speech. The original programs used hand-constructed phoneme detectors, but as computer power grew that work was entrusted to automatic learning. In 1992 Reddy's group demonstrated a program called Sphinx II on a 15-MIPS workstation with 100 MIPS of specialized signal-processing circuitry. Sphinx II was able to deal with arbitrary English speakers using a several-thousand-word vocabulary. The system's word detectors, encoded in statistical structures known as Markov tables, were shaped by an automatic learning process that digested hundreds of hours of spoken examples from thousands of Carnegie Mellon volunteers enticed by rewards of pizza and ice cream. Several practical voice-control and dictation systems are sold for personal computers today, and some heavy users are risking larynxes to save wrists.

More computer power is needed to reach human performance, but how much? Human and animal brain sizes imply an answer, if we can relate nerve volume to computation. Structurally and functionally, one of the best understood neural assemblies is the retina of the vertebrate eye. Happily, retina-like operations have been developed for robot vision, handing us a rough conversion factor.

The retina is a transparent, paper-thin layer of nerve tissue at the back of the eyeball on which the eye's lens projects an image of the world. It is connected by the optic nerve, a million-fiber cable, to regions deep in the brain. It is a part of the brain convenient for study, even in living animals, because of its peripheral location and because its function is straightforward compared with the brain's other mysteries. A human retina is about a centimeter across and a half-millimeter thick. It has about 100 million neurons, of five distinct kinds. Light-sensitive cells feed wide spanning *horizontal* cells and narrower *bipolar* cells, which are interconnected by *amacrine* cells, and finally *ganglion* cells, whose outgoing fibers bundle to form the optic nerve. Each of the million ganglion-cell axons carries signals from a particular patch of image, indicating light intensity differences over space or time: a million simultaneous edge and motion detections. Overall, the retina seems to process about ten images per second.

It takes robot vision programs about 100 computer instructions to derive single edge or motion detections from comparable video images. One hundred million instructions would be needed to do a million detections, and 1,000 MIPS to repeat them ten times per second to match the retina, a power just being reached by high-end personal computers.

The 1,500 cubic centimeter human brain is about one hundred thousand times as large as the retina, suggesting that matching overall human function will take about 100 *million* MIPS of computer power. Computer chess bolsters this yardstick. Deep Blue, the chess machine that bested world chess champion Garry Kasparov in 1997, used specialized chips to process chess moves at a rate equivalent to a 3 million MIPS universal computer (see figure on page 71). This is 3% of the estimate for total human performance. Since it is plausible that Kasparov, probably the best human player ever, can apply his brainpower to the strange problems of chess with an efficiency of 3%, Deep Blue's near parity with Kasparov's chess skill supports the retina-based extrapolation.

The most powerful experimental supercomputers in 1998, composed of thousands or tens of thousands of the fastest microprocessors and costing tens of millions of dollars, can do a few million MIPS. They are within striking distance of being powerful enough to match human brainpower, but they are unlikely to be applied to that end. Why tie up a rare twenty-million-dollar asset to develop one ersatz-human, when millions of inexpensive original-model humans are available? Such machines are needed for high-value scientific calculations, mostly physical simulations, having no cheaper substitutes. AI research must wait for the power to become more affordable.

If 100 million MIPS could do the job of the human brain's one hundred billion neurons, then one neuron is worth about 1/1,000 MIPS, that is, 1,000 instructions per second. That's probably not enough to simulate an actual neuron, which can produce 1,000 finely timed pulses per second. Our estimate is for very efficient programs that imitate the aggregate function of thousand-neuron assemblies. Almost all nervous systems contain subassemblies that big.

The small nervous systems of insects and other invertebrates seem to be hardwired from birth, with each neuron having its own special predetermined links and function. The few-hundred-million-bit insect genome is big enough to specify connections of each of their hundred thousand neurons. Humans, on the other hand, have one hundred billion neurons, but only a few billion bits in their genome, not enough to specially encode every connection. The human brain seems to consist largely of regular structures whose neurons are trimmed away as skills are learned, like featureless marble blocks chiseled into individual sculptures. Analogously, robot programs were precisely hand-coded when they occupied only a few hundred thousand bytes of memory. Now that they've grown to tens of millions of bytes, most of their content is learned from example. But there is a big practical difference between animal and robot learning. Animals learn individually, but robot learning can be copied from one machine to another. For instance, today's text and speech recognition programs were painstakingly trained over months or years, but each customer's copy of the software is "born" fully educated. Decoupling training from use will allow robots to do more with less. Big computers at the factory—maybe supercomputers a thousand times too big and expensive to fit in a robot—will process large training sets under careful human supervision and distill the results into efficient programs and arrays of settings that are then copied into the modest processors of myriads of individual robots.

Programs need memory as well as processing speed to do their work. The ratio of memory to speed has remained remarkably constant during computing history. The earliest electronic computers had a few thousand bytes of memory and could do a few thousand calculations per second. Medium computers of 1980 had a million bytes of memory and did a million calculations per second. Supercomputers in 1990 did a billion calculations per second and had a billion bytes of memory. The latest, greatest supercomputers can do a trillion calculations per second and can have a trillion bytes of memory. Dividing memory by speed defines a "time constant," roughly how long it takes the computer to scan once through its memory. One megabyte per MIPS gives one second, a nice human interval. Machines with less memory for their speed, typically new models, seem fast, but are unnecessarily limited to small programs. Models with more memory for their speed, often ones reaching the end of their commercial viability, can handle larger programs, but unpleasantly slowly. For instance, the original Macintosh was introduced in 1984 with 1/2 MIPS and 1/8 megabyte, and was then considered a very fast machine. The "fat Mac" with 1/2 MIPS and 1/2 megabyte ran larger programs at tolerable speed, but the 1/2-MIPS, 1-megabyte Mac+ verged on slow. The 4-megabyte Mac classic, the last 1/2-MIPS machine in the line, was intolerably slow and was soon supplanted by processors ten times faster in the same enclosure. Customers maintain the ratio by asking, "Would the next upgrade be better spent on more speed or more memory?"

The best evidence about nervous system memory puts most of it in the synapses connecting the neurons. Molecular adjustments allow synapses to be in a number of distinguishable states, let's say one byte's worth. Then the one-hundred-trillion-synapse human brain would hold the equivalent 100 million megabytes. Compare this to our earlier estimate that it would take 100 million MIPS to mimic the brain's function. The megabyte per MIPS rule seems to hold for nervous systems too! It is as if interactive machines were outboard continuations of our nervous systems, maintaining the same time constants. Machines that interact directly with the world rather than through people can have different speed to memory ratios. Fast-acting machines, for instance embedded audio and video processors and controllers of high-performance aircraft, have many MIPS for each megabyte. Very slow machines, for instance time-lapse security cameras and automatic data libraries, store many megabytes for each of their MIPS. Likewise, flying insects seem to be a few times faster than humans and so may have more MIPS than megabytes. As in animals, cells in plants signal one another electrochemically and enzymatically. Some plant cells seem specialized for communication, though apparently not as extremely as animal neurons. One day we may find that plants remember much, but process it slowly. (How does a redwood tree manage to rebuff rapidly evolving pests during a 2,000-year lifespan, when it took mosquitoes only a few decades to overcome DDT?)

With the conversions above, a 100-MIPS robot, for instance the last chapter's road-following Navlab, has mental power similar to a 100,000-neuron housefly. The following figure rates various entities on this scale.

Bicycle Race

By our estimate, today's very biggest supercomputers are within a factor of 100 of having the power to mimic a human mind. Their successors a decade hence will be more than powerful enough. Yet, it is unlikely that machines costing tens of millions of dollars will be wasted doing what any human can do, when they could instead be solving urgent physical and mathematical problems nothing else can touch. Machines with human-like performance will make economic sense only when they cost less than humans, say, when their "brains" cost about $1,000. When will that day arrive?

The expense of computation has fallen rapidly and persistently for a century. Steady improvements in mechanical and electromechanical calculators before World War II had increased the speed of calculation a thousandfold over hand calculation. The pace quickened with the appearance of electronic computers during the war—from 1940 to 1980 the amount of computation available at a given cost increased a millionfold. Vacuum tubes were replaced by transistors, and transistors by integrated circuits, whose components became ever smaller and more numerous. During the 1980s microcomputers reached the consumer market, and the industry became more diverse and competitive. Powerful, inexpensive computer workstations replaced the drafting boards of electronics and computer designers, and an increasing number of design steps were automated. The time to bring a new generation of computer to market shrank from two years at the beginning of the 1980s to less than nine months. The computer and communication industries grew into the largest on earth.

Computers doubled in capacity every two years after the war, a pace that became an industry given: companies that wished to grow sought to exceed it, companies that failed to keep up lost business. In the 1980s the doubling time contracted to eighteen months, and computer performance in the late 1990s seems to be doubling every twelve months.

MIPS and megabytes

*Entities rated by the computational power and memory of the smallest universal computer needed to mimic their behavior. Note that the scale is logarithmic on both axes: each vertical division represents a thousandfold increase in processing power, and each horizontal division a thousandfold increase in memory size. Universal computers, marked by an *, can imitate other entities at their location in the diagram, but the more specialized entities cannot. A 100-million-MIPS computer may be programmed not only to think like a human, but also to imitate other similarly sized computers. But humans cannot imitate 100-million-MIPS computers—our general-purpose calculation ability is under a millionth of a MIPS. Deep Blue's special-purpose chips process chess moves like a 3-million-MIPS computer, but its general-purpose power is only a thousand MIPS. Most of the noncomputer entities in the diagram can't function in a general-purpose way at all. Universality is an almost magical property, but it has costs. A universal machine may use ten or more times the resources of one specialized for a task. But if the task should change, as it usually does in research, the universal machine can be reprogrammed, while the specialized machine must be replaced.*

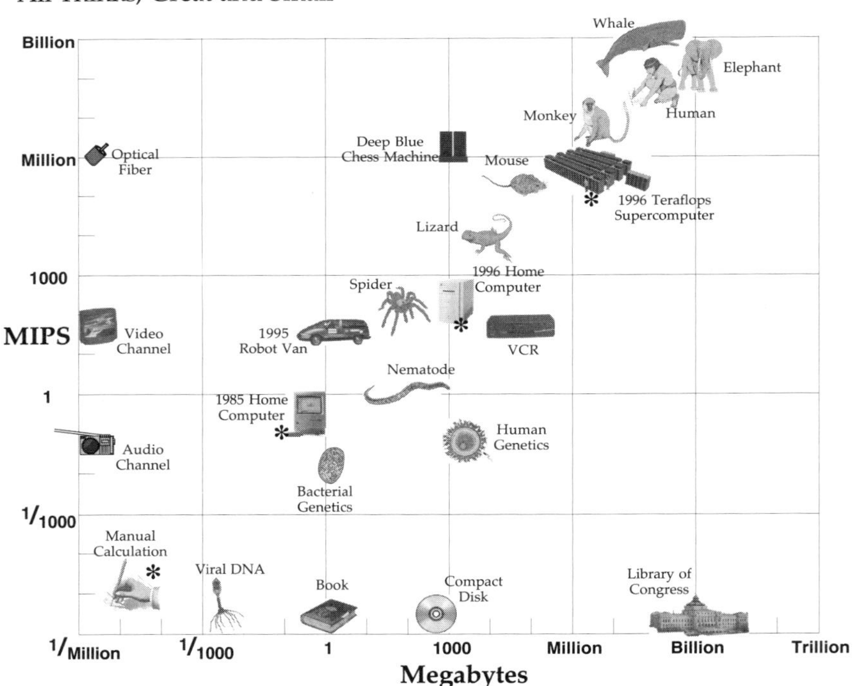

At the present rate, computers suitable for humanlike robots will appear in the 2020s. Can the pace be sustained for another three decades? The rate shows no sign of abatement. If anything, it hints that further contractions in timescale are in store. But, one often encounters thoughtful articles by knowledgeable people in the semiconductor industry giving detailed reasons why the decades of phenomenal growth must soon come to an end.

The keynote for advancing computation is miniaturization: smaller components have less inertia and operate more quickly with less energy, and more of them can be packed in a given space. First the moving parts shrunk, from the gears in mechanical calculators, to small contacts in electromechanical machines, to bunches of electrons in electronic computers. Next, the switches' supporting structure underwent a vanishing act, from thumb-sized vacuum tubes, to fly-sized transistors, to ever-diminishing flyspecks on integrated circuit chips. Similar to printed circuits before them, integrated circuits were made by a photographic process. The desired pattern was projected onto a silicon chip, and subtle chemistry was used to add or remove the right sorts of matter in the exposed areas.

In the mid-1970s, integrated circuits, age fifteen, hit a crisis of adolescence. They then held ten thousand components, just enough for an entire computer, and their finest details were approaching 3 micrometers in size. Experienced engineers wrote many articles warning that the end was near. Three micrometers was barely larger than the wavelength of the light used to sculpt the chip. The number of impurity atoms defining the tiny components had grown so small that statistical scatter would soon render most components out of spec, a problem aggravated by a similar effect in the diminishing number of signaling electrons. Increasing electrical gradients across diminishing gaps caused atoms to creep through the crystal, degrading the circuit. Interactions between ever-closer wires were about to ruin the signals. Chips would soon generate too much heat to remove, and require too many external connections to fit. The smaller memory cells were suffering radiation-induced forgetfulness.

A look at the following computer growth graph shows that the problems were overcome, with a vengeance. Chip progress not only continued, it sped up. Shorter-wavelength light was substituted, a

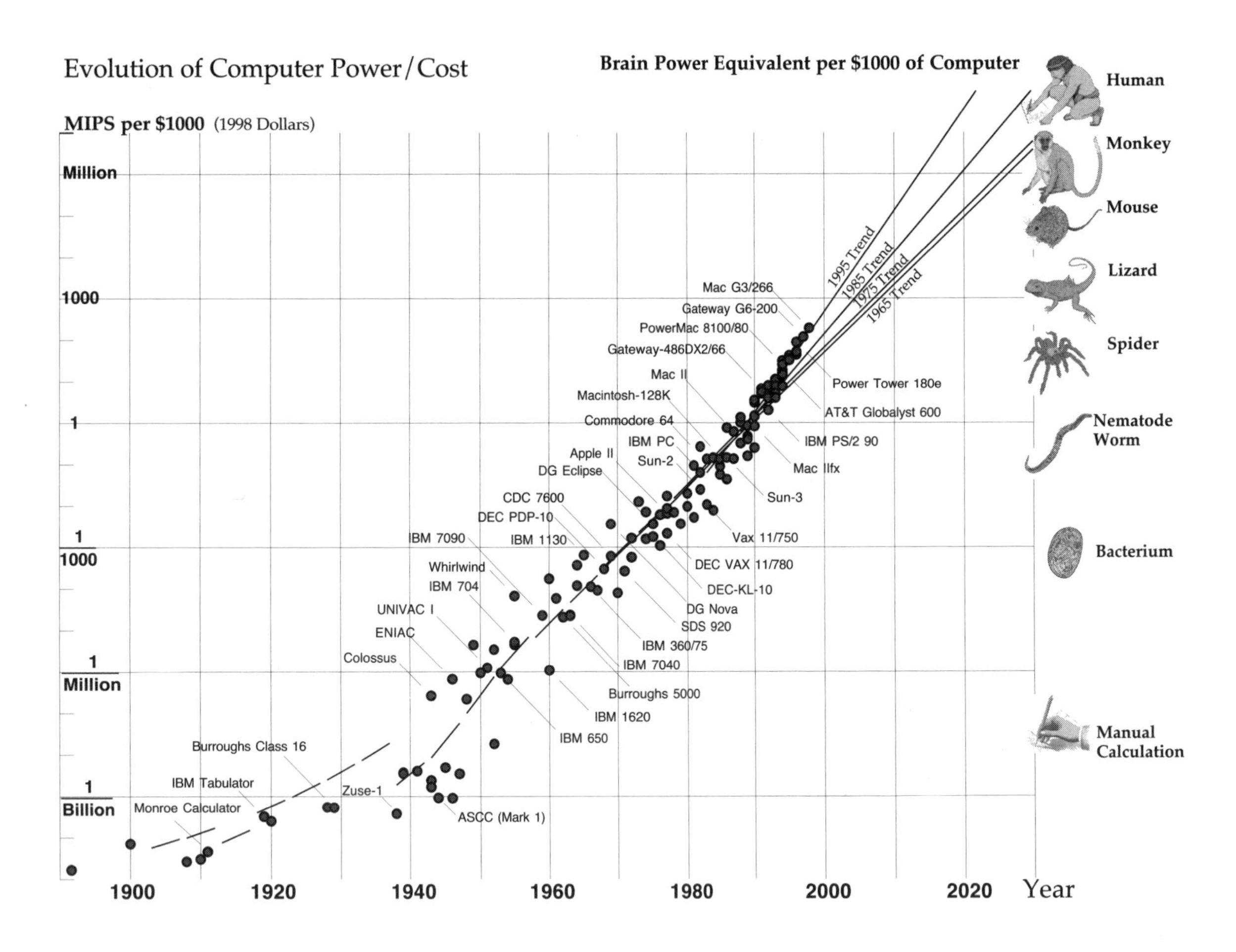

Evolution of Computer Power/Cost
Brain Power Equivalent per $1000 of Computer
MIPS per $1000 (1998 Dollars)
Million
1000
1
1
1000
1
Million
1
Billion
1900
1920
1940
1960
1980
2000
2020
Year
Human
Monkey
Mouse
Lizard
Spider
Nematode Worm
Bacterium
Manual Calculation
1995 Trend
1985 Trend
1975 Trend
1965 Trend
Mac G3/266
Gateway G6-200
PowerMac 8100/80
Gateway-486DX2/66
Mac II
Macintosh-128K
Commodore 64
IBM PC
Apple II
Sun-2
DG Eclipse
CDC 7600
DEC PDP-10
IBM 1130
IBM 7090
Whirlwind
IBM 704
UNIVAC I
ENIAC
Colossus
Burroughs Class 16
IBM Tabulator
Monroe Calculator
Zuse-1
ASCC (Mark 1)
Power Tower 180e
AT&T Globalyst 600
IBM PS/2 90
Mac IIfx
Sun-3
Vax 11/750
DEC VAX 11/780
DEC-KL-10
DG Nova
SDS 920
IBM 360/75
IBM 7040
Burroughs 5000
IBM 1620
IBM 650

← **Faster than exponential growth in computing power**
The number of MIPS in $1,000 of computer from 1900 to the present. Steady improvements in mechanical and electromechanical calculators before World War II had increased the speed of calculation a thousandfold over manual methods from 1900 to 1940. The pace quickened with the appearance of electronic computers during the war, and 1940 to 1980 saw a millionfold increase. Since then the pace has been even quicker, a pace that would make humanlike robots possible before the middle of the next century. The vertical scale is logarithmic; the major divisions represent thousandfold increases in computer performance. Exponential growth would show as a straight line, the upward curve indicates faster than exponential growth, an accelerating rate of innovation. The reduced spread of the data in the 1990s is probably the result of intensified competition: underperforming machines are more rapidly squeezed out.

---

more precise way of implanting impurities was devised, voltages were reduced, better insulators, shielding designs, more efficient transistor designs, better heat sinks, denser pin patterns, and non-radioactive packaging materials were found. Where there is sufficient financial incentive, there is a way. In fact, solutions had been waiting in research labs for years, barely noticed by the engineers in the field, who were perfecting established processes and worrying in print as those ran out of steam. As the need became acute, enormous resources were redirected to draft laboratory possibilities into production realities.

The intervening years saw many additional problems, solutions, and innovations, but now, nearing a midlife forty, the anxieties seem again to be cresting. In 1996 major articles appeared in scientific magazines and major national newspapers worrying that electronics progress might be within a decade of ending. The cost of building new integrated circuit plants was approaching a prohibitive billion dollars. Feature sizes were reaching 0.1 micrometers, the wavelength of the sculpting ultraviolet light. Integrated transistors, scaled down steadily from 1970s designs, would soon be so small that electrons would quantum "tunnel" out of them. Wiring was becoming so dense it would crowd out the components, and slow down and leak signals. Heat was increasing.

The articles didn't mention that less expensive plants could make the same integrated circuits, if less cheaply and in smaller quantities. Scale was necessary because the industry had grown so large and competitive. Instead of signaling impending doom, it indicated free-market success, a battle of titans driving down costs to the users. They also failed to mention new contenders, waiting on lab benches to step in, should the leading techniques falter.

The wavelike nature of matter at very small scales is a problem for conventional transistors, which depend on the smooth flow of masses of electrons. But it is a property exploited by a radical new class of components known as single-electron transistors and quantum dots, which work by the interference of electron waves. These new devices work better as they grow smaller. At the scale of today's circuits, the interference patterns are so fine that it takes only a little heat energy to bump electrons from crest to crest, scrambling their operation. Thus, these circuits have been demonstrated mostly at a few degrees above absolute zero. But, as the devices are reduced, the interference patterns widen, and it takes ever larger energy to disrupt them. Scaled to about 0.01 micrometers, quantum interference switching works at room temperature. It promises more than a thousand times higher density than today's circuits, possibly a thousand times the speed, and much lower power consumption, since it moves a few electrons across small quantum bumps, rather than pushing them in large masses through resistive material. In place of much wiring, quantum interference logic may use chains of switching devices. It could be manufactured by advanced descendants of today's chip-fabrication machinery. Proposals abound in the research literature,[9] and the industry has the resources to perfect the circuits and their manufacture when the time comes.

Wilder possibilities are brewing. Switches and memory cells made of single molecules have been demonstrated, which might enable a volume to hold a billion times more circuitry than today. Potentially blowing everything else away are "quantum computers," in which a whole computer, not just individual signals, acts in a wavelike manner. Like a conventional computer, a quantum computer consists of a number of memory cells whose contents are modified in a sequence of logical transformations. Unlike a conventional computer, whose memory cells are either 1 or 0, each cell in a quantum computer is started in a quantum superposition of both 1 and 0. The whole machine is a superposition of all possible combinations of memory states. As the computation proceeds, each component of the superposition individually undergoes the logic operations. It is as if an exponential number of computers, each starting with a different pattern in memory, were working on the problem simultaneously. When the computation is finished, the memory cells are examined, and an answer emerges from the wavelike interference of all the possibilities. The trick is to devise the computation so that the desired answers reinforce, while the others cancel. In the last several years, quantum algorithms have been devised that factor numbers and search for encryption keys much faster than any classical computer. Toy quantum computers, with three or four "qubits" stored as states of single atoms or photons, have been demonstrated, but they can do only short computations before their delicate superpositions are scrambled by outside interactions. More promising are computers using nuclear magnetic resonance, as in hospital scanners. There, quantum bits are encoded as the spins of atomic nuclei, and gently nudged by external magnetic and radio fields into magnetic interactions with neighboring nuclei. The heavy nuclei, swaddled in diffuse orbiting electron clouds, can maintain their quantum coherence for hours or longer. A quantum computer with a thousand or more qubits could tackle problems astronomically beyond the reach of any conceivable classical computer.

Molecular and quantum computers will be important sooner or later, but humanlike robots are likely to arrive without their help. Research within semiconductor companies, including working prototype chips, makes it quite clear that existing techniques can be nursed along for another decade, to chip features below 0.1 micrometers, memory chips with tens of billions of bits and multiprocessor chips with over 100,000 MIPS. Toward the end of that period, the circuitry will probably incorporate a growing number of quantum interference components. As production techniques for those tiny components are perfected, they will begin to take over the chips, and the pace of computer progress may steepen further. The 100 million MIPS to match human brain power will then arrive in home computers before 2030.

The next chapter lays out an incremental development schedule for the software of intelligent robots, following this anticipated hardware pace. In it, robot minds evolve in stages roughly paralleling the evolution of our own brains, but ten million times as fast, reaching humanlike intelligence in about forty years.

False Start

It may seem rash to expect fully intelligent machines in a few decades, when the computers have barely matched insect mentality in a half-century of development. Indeed, for that reason, many long-time artificial intelligence researchers scoff at the suggestion, and offer a few centuries as a more believable period. But there are very good reasons why things will go much faster in the next fifty years than they have in the last fifty.

The stupendous growth and competitiveness of the computer industry are one reason. A less-appreciated one is that intelligent machine research did not make steady progress in its first fifty years—it marked time for thirty of them! Although general computer power grew a hundred thousandfold from 1960 to 1990, the computer power available to AI programs barely budged from 1 MIPS during those three decades.

In the 1950s the pioneers of AI viewed computers as locomotives of thought, which might outperform humans in higher mental work as prodigiously as they outperformed them in arithmetic, if they were harnessed to the right programs. Success in the endeavor would bring enormous benefits to national defense, commerce, and government. The promise warranted significant public and private investment. For instance, there was a large project to develop machines to automatically translate scientific and other literature from Russian to English. There were only a few AI centers, but those had the largest computers of the day, comparable in cost to today's supercomputers. A common one was the IBM 704, which provided a good fraction of a MIPS.

By 1960 the unspectacular performance of the first reasoning and translation programs had taken the bloom off the rose, but the unexpected launching by the Soviet Union of Sputnik, the first satellite, in 1957, had substituted a paranoia. Artificial Intelligence may not have delivered on its first promise, but what if it were to suddenly succeed after all? To avoid another nasty technological surprise from the enemy, it behooved the United States to support the work, moderately, just in case. Moderation paid for medium-scale machines costing a few million dollars, no longer supercomputers. In the 1960s that price provided a good fraction of a MIPS in thrifty machines like Digital Equipment Corp's innovative PDP-1 and PDP-6.

The field looked even less promising by 1970, and support for military-related research declined sharply with the end of the Vietnam War. Artificial Intelligence research was forced to tighten its belt and beg for unaccustomed small grants and contracts from science agencies and industry. The major research centers survived, but became a little shabby as they made do with aging equipment. For almost the entire decade AI research was done with PDP-10 computers that provided just under 1 MIPS. Because it had contributed to the design, the Stanford AI Lab received a 1.5-MIPS KL-10 in the late 1970s from Digital, as a gift.

Funding improved somewhat in the early 1980s, but the number of research groups had grown, and the amount available for computers was modest. Many groups purchased Digital's new Vax computers, costing $100,000 and providing 1 MIPS. By mid-decade, personal computer workstations had appeared. Individual researchers reveled in the luxury of having their own computers, avoiding the delays of time-shared machines. A typical workstation was a Sun-3, costing about $10,000, and providing about 1 MIPS.

By 1990 entire careers had passed in the frozen winter of 1-MIPS computers, mainly from necessity, but partly from habit and a lingering opinion that the early machines really should have been powerful enough. In 1990, 1 MIPS cost $1,000 in a low-end personal computer. There was no need to go any lower. Finally spring thaw has come. Since 1990, the power available to individual AI and robotics programs has doubled yearly, to 30 MIPS by 1994 and 500 MIPS by 1998. Seeds long ago alleged barren are suddenly sprouting. Machines read text, recognize speech, even translate languages. Robots drive cross-country, crawl across Mars, and trundle down office corridors. In 1996 a theorem-proving program called EQP running five weeks on a 50-MIPS computer at Argonne National Laboratory found a proof of a Boolean algebra conjecture by

Herbert Robbins[10] that had eluded mathematicians for sixty years. Just as impressive, if harder to quantify, is the music composition program EMI, an ongoing project by David Cope. EMI distills essential patterns from bodies of work by other composers and produces new music in their style—i.e., it learns from them. In recent years EMI's classical compositions have pleased audiences, who rate it above most human composers. And it is still only spring. Wait until summer.

The Game's Afoot

A summerlike air already pervades the few applications of artificial intelligence that retained access to the largest computers. Some of these, like pattern analysis for satellite images and other kinds of spying, and for seismic oil exploration, are closely held secrets. Another, though, basks in the limelight. The best chess-playing computers are so interesting that they generate millions of dollars of free advertising for the winners, and consequently they have enticed a series of computer companies to donate time on their best machines along with other resources. Since 1960 IBM, Control Data, AT&T, Cray, Intel, and now again IBM have been sponsors of computer chess. The "knights" in the following AI power graph show the effect of this largess, relative to mainstream AI research. The top chess programs have competed in tournaments powered by supercomputers or specialized machines whose chess power is comparable. In 1958 IBM had both the first checker program, by Arthur Samuel, and the first full chess program, by Alex Bernstein. They ran on an IBM 704, the biggest and last vacuum-tube computer. The Bernstein program played atrociously, but Samuel's program, which automatically learned its board-scoring parameters, was able to beat Connecticut checkers champion Robert Nealey. Since 1994, Chinook, a program written by Jonathan Schaeffer of the University of Alberta, has consistently bested the world's human checker champion. But checkers isn't very glamorous, and this portent received little notice.

By contrast, it was nearly impossible to overlook the epic battles between world chess champion Garry Kasparov and IBM's Deep Blue in 1996 and 1997. Deep Blue is a scaled-up version of a machine called Deep Thought, built by Carnegie Mellon University students ten years earlier. Deep Thought, in turn, depended on special-purpose chips, each wired like the Belle chess computer built by Ken Thompson at AT&T Bell Labs in the 1970s. Belle, organized like a chessboard, circuitry on the squares, wires running like chess moves, could evaluate a position and find all legal moves from it in one electronic flash. In 1997 Deep Blue had 256 such chips, orchestrated by a 32-processor mini-supercomputer. It examined 200 million chess positions a second. Chess programs, on unaided general-purpose computers, average about 16,000 instructions per position examined. Deep Blue, when playing chess (and only then), was thus worth about 3 million MIPS, one thirtieth of our estimate for human intelligence, or the full brainpower of a small monkey.

Deep Blue, in a first for machinekind, won the first game of the 1996 match. But, Kasparov quickly found the machine's weaknesses, and drew two and won three of the remaining games.

In May 1997 he met an improved version of the machine. That February, Kasparov had triumphed over a field of grandmasters in a prestigious tournament in Linares, Spain, reinforcing his reputation as the best player ever, and boosting his chess rating past 2800—uncharted territory. He prepared for the computer match in the intervening months, in part by playing against other machines. Kasparov won a long first game against Deep Blue, but lost the next day to masterly moves by the machine. Then came three grueling draws and a final game, in which a visibly shaken and angry Kasparov resigned early, with a weak position. The 1997 contest with Deep Blue was the first competition match Kasparov had ever lost.

The event was notable for many reasons, but one especially is of interest here. Several times during both matches, Kasparov reported signs of mind in the machine. In the second tournament, he worried there might be humans behind the scenes, feeding Deep Blue strategic insights!

Bobby Fischer, the U.S. chess great of the 1970s, is reputed to have played each game as if against God, simply making the best moves. Kasparov, on the other hand, claims to see into opponents' minds during play, intuiting and exploiting their plans, insights, and oversights. In ordinary chess computers, he reports a mechan-

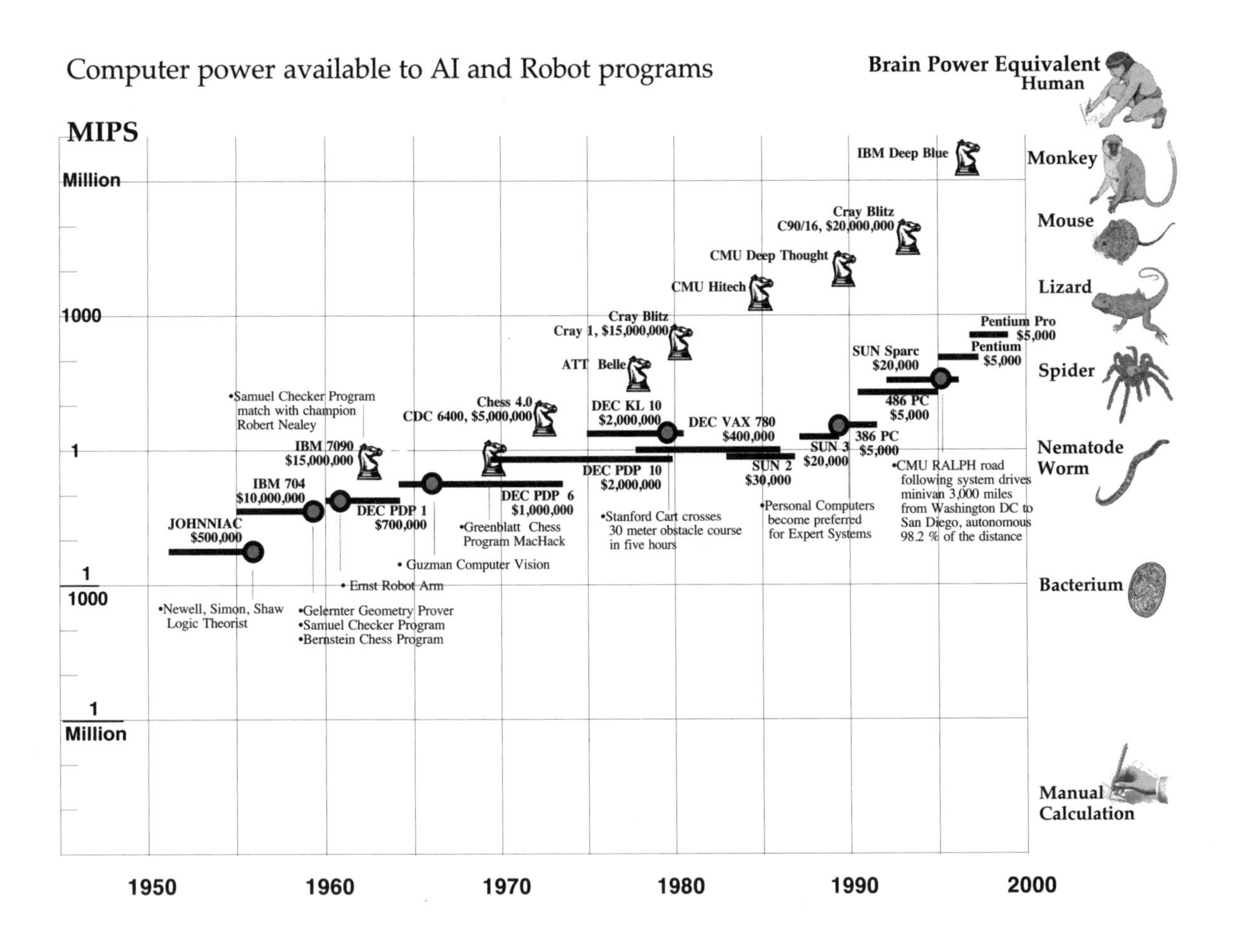

Computer power available to AI and Robot programs
Brain Power Equivalent
Human
MIPS
Million
1000
1
1
1000
1
Million
1950
1960
1970
1980
1990
2000
Monkey
Mouse
Lizard
Spider
Nematode Worm
Bacterium
Manual Calculation
IBM Deep Blue
Cray Blitz
C90/16, $20,000,000
CMU Deep Thought
CMU Hitech
Cray Blitz
Cray 1, $15,000,000
ATT Belle
Pentium Pro
$5,000
Pentium
$5,000
SUN Sparc
$20,000
486 PC
$5,000
386 PC
$5,000
SUN 3
$20,000
SUN 2
$30,000
DEC VAX 780
$400,000
DEC KL 10
$2,000,000
DEC PDP 10
$2,000,000
Chess 4.0
CDC 6400, $5,000,000
DEC PDP 6
$1,000,000
DEC PDP 1
$700,000
IBM 7090
$15,000,000
IBM 704
$10,000,000
JOHNNIAC
$500,000
•Samuel Checker Program match with champion Robert Nealey
•Greenblatt Chess Program MacHack
• Guzman Computer Vision
• Ernst Robot Arm
•Newell, Simon, Shaw Logic Theorist
•Gelernter Geometry Prover
•Samuel Checker Program
•Bernstein Chess Program
•Stanford Cart crosses 30 meter obstacle course in five hours
•Personal Computers become preferred for Expert Systems
•CMU RALPH road following system drives minivan 3,000 miles from Washington DC to San Diego, autonomous 98.2 % of the distance

← **The big freeze**
From 1960 to 1990 the cost of computers used in AI research declined while their numbers increased as funding decreased. The dilution absorbed computer-efficiency gains during the period, and the power available to individual AI programs remained almost unchanged at 1 MIPS—less than insect power. AI computer cost bottomed in 1990, and since then power has doubled yearly, to several hundred MIPS by 1998. The major visible exception to this pattern is computer chess, shown by a progression of knights, whose prestige lured the resources of major computer companies and the talents of programmers and machine designers. Exceptions also exist in less public competitions, like petroleum exploration and intelligence gathering, whose high return on investment warrants access to the largest computers.

---

ical predictability, stemming from their undiscriminating but limited lookahead and absence of long-term strategy. In Deep Blue, to his consternation, he saw instead an "alien intelligence."

In this paper-thin slice of mentality, a computer seems to have not only outperformed the best human, but transcended its machinehood. Who better to judge than Garry Kasparov? Mathematicians who examined EQP's proof of the Robbins conjecture, mentioned earlier, report a similar impression of creativity and intelligence, as do music connoisseurs in some of EMI's compositions. In these cases, the evidence for an intelligent mind lies in the machine's performance, not its makeup.

Now, the team that built Deep Blue claim no "intelligence" for it, only a large database of opening and end games, scoring and deepening functions tuned with the advice of consulting grandmasters, and, especially, raw speed that allows the machine to look ahead an average of fourteen half-moves per turn. Unlike some earlier, less successful, chess programs, Deep Blue was not designed to think like a human, to form abstract strategies or see patterns as it races through the move/countermove tree as fast as possible.

Deep Blue's creators know its *quantitative* superiority over other chess machines intimately, but lack the chess understanding to share Kasparov's deep appreciation of the difference in the *quality* of its play. I think this dichotomy will show up increasingly in the coming years. Engineers who know the mechanism of advanced robots most intimately will be the last to admit they have real minds. From the inside, robots will indisputably be machines, acting according to mechanical principles, however elaborately layered. Only on the outside, where they can be appreciated as a whole, will the impression of intelligence emerge. A human brain, too, does not exhibit the intelligence under a neurobiologist's microscope that it does participating in a lively conversation.

The Great Flood

Computers are universal machines; their potential extends uniformly over a boundless expanse of tasks. Human potentials, on the other hand, are strong in areas long important for survival, but weak in things far removed. Imagine a "landscape of human competence," having lowlands with labels like "arithmetic" and "rote memorization," foothills like "theorem proving" and "chess playing," and high mountain peaks labeled "locomotion," "hand-eye coordination," and "social interaction." We all live in the solid mountaintops, but it takes great effort to reach the rest of the terrain, and only a few of us work each lowland patch.

Advancing computer performance is like water slowly flooding the landscape. A half-century ago it began to drown the lowlands, driving out human calculators and record clerks, but leaving most of us dry. Now the flood has reached the foothills, and our outposts there are contemplating retreat. We feel safe on our peaks, but, at the present rate, those too will be submerged within another half-century. Chapter 5 proposes that we build arks as that day nears and adopt a seafaring life, lifted rather than drowned by the flood! For now, though, we must rely on our representatives in the lowlands to tell us what the water is really like.

Waterworld

Our representatives on the foothills of chess and theorem proving report signs of intelligence. Why didn't we get similar reports decades before, from the lowlands, as computers surpassed humans in arithmetic and rote memorization? Actually, we did, at the time. Computers that calculated like thousands of mathematicians were hailed as "giant brains" and inspired the first generation of AI research. After all, the machines were doing something beyond

Agony to ecstasy

In forty years, computer chess progressed from the lowest depth to the highest peak of human chess performance. It took a handful of good ideas, culled by trial and error from a larger number of possibilities, an accumulation of previously evaluated game openings and endings, good adjustment of position scores, and especially a ten-millionfold increase in the number of alternative move sequences the machines can explore. Note that chess machines reached world champion performance as their (specialized) processing power reached about 3% of human, on our brain-to-computer scale. Since it is plausible that Garry Kasparov (but hardly anyone else) can apply his brainpower to the problems of chess with an efficiency of 3%, the result supports our retina-based extrapolation. In coming decades, as general-purpose computer power grows beyond Deep Blue's specialized strength, machines will begin to match humans in more common skills.

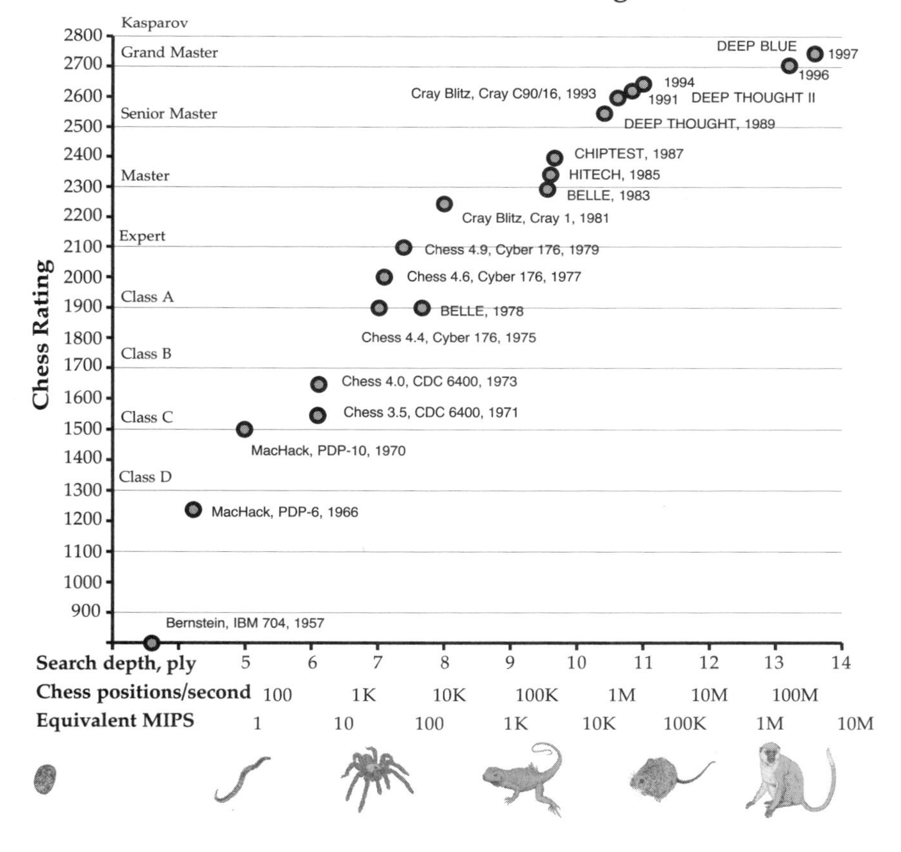

any animal, which needed human intelligence, concentration, and years of training. But it is hard to recapture that magic now. One reason is that computers' demonstrated stupidity in other areas biases our judgment. Another relates to our own ineptitude. We do arithmetic or keep records so painstakingly and externally that the small mechanical steps in a long calculation are obvious, while the big picture often escapes us. Like Deep Blue's builders, we see the process too much from the inside to appreciate the subtlety that it may have on the outside. But there is a nonobviousness in snowstorms or tornadoes that emerge from the repetitive arithmetic of weather simulations, or in rippling tyrannosaur skin from movie animation calculations. We rarely call it intelligence, but "artificial reality" may be an even more profound concept than artificial intelligence. More on that in Chapter 7.

The mental steps underlying good human chess playing, theorem proving, or music composition are complex and hidden, putting a mechanical interpretation out of reach. Those who can follow the play naturally describe it in mentalistic language, using terms like strategy, understanding, creativity, beauty, or emotion. When a machine manages to be simultaneously meaningful and surprising in the same rich way, it too compels a mentalistic interpretation. Of course, somewhere behind the scenes there are programmers who, in principle, have a mechanical interpretation. But even for them, that interpretation loses its grip as the working program fills its memory with details too voluminous and complex for them to grasp.

As the rising flood reaches more populated heights, machines will begin to do well in areas a greater number of us can appreciate. The visceral sense of a thinking presence in machinery will become increasingly widespread. When the highest peaks are covered, there will be machines that can interact as intelligently as any human on any subject. The presence of minds in machines will then become self-evident.

Alan Turing anticipated such a development a half-century ago.

Turing Tests

In 1949 at Manchester University, site of the first working general-purpose digital computer, Michael Polanyi organized a symposium on the subject *The Mind and the Computing Machine.* Interested philosophers, biologists, psychologists, mathematicians, physicists, and others from all Britain contributed to an intense discussion that presaged the thinking machines debate for the rest of the century. At the center was Alan Turing, whose vision had shaped the Manchester computer, as well as secret wartime code-breaking machines, and who had contemplated the mechanization of thought for two decades. Turing summarized his responses to the major arguments against machine intelligence in a now-famous paper, *Computing Machinery and Intelligence*, in a 1950 issue of the philosophical journal *Mind*.

In the paper, Turing observed that the terms *thinking* and *machinery* were too vague for a precise answer to the question "Can Machines Think?" Would a biological, but artificially constructed, human being count as a machine? Did solving difficult arithmetic problems count as thinking? He suggested substituting a less ambiguous question. A human judge engages in conversations by teletype with both an intelligent human and a specially programmed digital computer. If, after a certain length of freewheeling conversation, the judge is, as often as not, fooled into mistaking the machine for the human, the machine can be declared an intelligent thinker. Turing guessed that by the year 2000 computers with a billion bytes of memory could be programmed to pass five-minute tests. The prediction is still marginally viable in 1998, when machines with a billion bytes of disk storage are ubiquitous, machines with a billion bytes of main memory many, and programs are starting to pass "restricted topic" versions of the test. But the magnitude of the problem is better appreciated now, and it seems likely Turing's prediction will come fully true only a few decades hence, on machines a thousand times as powerful as his guess.

Although he had ideas for how to approach the problem, Turing was unable to offer an overwhelming argument that machines could be made to think. Instead, the paper rebutted opposing positions, grouped into the following nine categories:

1. The Theological Objection—*Thinking is a function of the soul. Machines have no souls, so cannot think.*

2. The "Heads in the Sand" Objection—*Thinking machines cannot be possible, because the consequences would be too dreadful.*

3. The Mathematical Objection—*Mechanical reasoning has certain provable limitations that human thought may not share.*

4. The Argument from Consciousness—*Machines have no inner experience to give meaning to their utterances, actions, or internal operations.*

5. Arguments from Various Disabilities—*Machines will never be kind, moral, joyous, perceptive, original, etc.*

6. Lady Lovelace's Objection—*Computers do only what we can program them to do.*

7. The Argument from Continuity in the Nervous System—*Nerves respond to arbitrarily tiny signal differences, while computers work in fixed-size steps.*

8. The Argument from Informality of Behavior—*It is not possible to specify for a machine what to do in every possible circumstance a human can encounter.*

9. The Argument from Extrasensory Perception—*Humans sometimes sense remote or future information unavailable to deterministic processes in computers.*

Computer power and numbers have grown astronomically since 1950, and the effort to elicit intelligence has become a major academic and commercial enterprise. Meanwhile, reputations have been built on public rebuttals of the idea of humanlike mechanical minds. Although the "pro" position has gained ground, neither side has yet scored a decisive victory, and the debate continues within the old framework. Perhaps, as Turing anticipated, perceptions will change when machines that read, understand, speak conversationally, and interact physically become common. Or perhaps the machines will simply continue the debate where flesh and blood left off! Here is a current cast on the questions and Turing's positions.

1 — The Theological Objection

Although an iconoclastic atheist, Turing could not escape England's state religion and its central dogma, the uniqueness and superiority of man, made in the image of God, with an immortal soul. A century before, Darwin's evolutionary theory had evoked outrage and ridicule in the best of British society because it suggested humans were a kind of animal. Turing's insinuations that the human mind was a clanking machine were simply over the top! Fierce nods sweep cultured British audiences when a speaker calls the idea "appalling"—it takes little imagination to hear growls of aggression. And why not? Our survival has depended on tribal integrity since before we were human, and we have powerful instincts to protect it. Religion is a cultural edifice built on those instincts, claiming a transcendent authority to guard tribal standards across generations.

Science seeks objective interpretations of observations, independent of human feelings, tribal values, and even its own traditions. Its mercurial course often subverts religion's role as social conservator, contradicting religious tenets and creating disturbing new options. Yet, despite a demonstrated potential for societal disruption, science has increasingly usurped religion's ancient explanations and rules because its material benefits outweighed the costs in peace of mind and social order. Turing noted that centuries of church opposition had not overturned Galileo's observations relegating the earth from the center of the universe to an unexceptional body moving among others. He might also have mentioned Darwinian evolution, whose interlocking evidence for our origin in ordinary, mindless processes explains and makes possible far more than do exalting stories of divine creation. Explanatory failures and the contradictions between different religions led Turing to dismiss religious claims. For the sake of argument, however, he asked if God had not the power to grant a soul to a machine constructed to host one? Fifty years later, the religious objection has become even less plausible, but perhaps there are approaches to the question that are fun even for atheists.

Religions have embellished it, but the concept of soul starts with the subjective sense of consciousness—of being who we are inside our bodies. The mechanistic idea that human consciousness arises solely from physical events in the brain and body seems, at first blush, diametrically opposed to the ancient view of life and mind as immaterial animating spirits. The contradiction may be only a difference of perspective. Not even the staunchest materialist claims that the chemical squirts of nerve cells are in themselves thoughts and feelings, rather only that large-scale patterns of brain activity can be interpreted as abstract mental events. Our minds make that interpretation—the very minds woven of those abstractions! No presumption of spirit is required to find a natural duality between the physical body and the abstract, self-interpreting mind.

"Number" is a simpler abstraction. Arranging abacus beads suggests, but does not create, numbers, and clearing the abacus does not destroy them. The numbers persist, appearing in other representations, minds, and logical implications. Taking the idea a little farther, the abacus, and the physical world in general, is just a large shared abstraction where small, independently existing abstractions like minds and numbers correlate. On this train of thought, there is every reason to believe that abstractions like minds and souls, no less than numbers, exist independently of their occasional representation in a physical world. Death of a body should no more destroy a soul—or its history or potential—than clearing an abacus destroys a number. Nor should death destroy sensations and consciousness—those are properties of the abstraction. Only the perfect correlation between the consciousness and the physical world would be lost.

So how does a machine get a soul? We grant a number to an abacus when we interpret the arrangement of beads as expressing that number. In the same way, we might grant a conscious soul to a robot by interpreting its behavior as expressing the action of such a soul: the more humanlike its interaction with us, the easier the attribution. Not everyone will agree. Social ethicists may interpret the robot's activity as only a parody of human behavior because the machine lacks the proper external links in its creation, history, or relationships—for instance, it may have been built with adult intelligence, never experiencing birth and growth. Functional mechanists may object that the robot's structural internals might not represent thoughts and feelings—for instance, what if it were controlled by simple (if huge) lookup tables with prearranged responses for every possible input sequence? Theologians and even some cosmologists may demand the imprimatur of a higher authority. Most people, however, are likely, after a period of suspicion, to begin taking machines that interact like intelligent, decent persons at face value, regardless of unseen internals, because it is the most effective alternative. Friendly people will greet robots with questions such as "How are you feeling today?" and receive answers such as "Terrible. I have a pain in all the diodes on my left side" or "Fine. I have the greatest enthusiasm for this mission." So, it may be appropriate to say "God" has granted a soul to a machine when the machine is accepted as a real person by a wide human community.

2 — The "Heads in the Sand" Objection

Turing saw some arguments against machine intelligence as rationalizations of fear of being displaced from the apex of things—and offered not rebuttal, but consolation "perhaps in the transmigration of souls." Why does an atheist suggest reincarnation? I think Turing was cryptically raising the possibility of transplanting human minds into other hardware, for instance, by converting essential brain functions to analogous computer programs, and body functions to robotic hardware. Such transplants might allow human personalities to transcend their biological limitations and join the machines' rapid evolution.

Among many nice reviews of my first book *Mind Children,* which outlined such a possibility, an angry few brandished words like "horrific," "nightmare," and "immoral," and at least one was too irate to publish. Intelligent machines may incite instinctive fear and anger by resembling ancestral threats—another tribe poaching in our territory, a rival for our social position, or a predator abducting our offspring. But thinking robots are none of the above; they are an entirely new kind of life. In behavior, robots resemble ourselves more than they resemble anything else in the world. They are being taught our skills. In the future they will acquire our values and goals; for instance, antisocial robot software would sell poorly, so most robots will behave in a decent way. How should we feel about beings that we bring into the world, that are similar to ourselves, that we teach our way of life, that will probably inherit the world when we are gone? I think we should consider them our children, a hope rather than a threat, though they will require careful upbringing to instill a good character. In time, they will outgrow us, create their own goals, make their own mistakes, and go their own way, with us perhaps a fond memory—but that too is the way of children.

3 — *The Mathematical Objection*

Large-brained organisms visualize possible futures in detailed time sequence, but humans can abandon details and time to shortcut through webs of contingency in the process called abstract reason. Unaided reason is limited by our small, transient, and unreliable short-term memories, and it was vastly enhanced when Aristotle and later thinkers expressed the process on paper. Inhumanly long, precise chains of deduction, leading from obvious, simple postulates to surprising, wonderful theorems, as in Euclid's geometry book, gave reason a mystical aura. The mystique grew when Descartes showed how to do geometry in arithmetic via graph paper, when later mathematicians derived arithmetic from handfuls of simple axioms, and when yet others created mathematical logics to encompass all thinking. In 1900 David Hilbert proposed a program to systematize all mathematics by deriving it in a formal, "mechanical" way from a few dozen axioms about sets, expecting that every truth could be cranked out eventually.

In 1931 Kurt Gödel showed how to encode arithmetical statements as huge numbers, allowing arithmetic's theorems, which heretofore had been interpreted as speaking only about numbers, to be seen as talking about themselves. With this interpretation, he then constructed a precise arithmetical statement meaning "*This statement cannot be deduced from the axioms of arithmetic.*" Now, if arithmetic contained no subtle contradictions, and so could not call itself a liar, Gödel's sentence could not be deduced. But if the sentence could not be deduced, then what it said was true! Gödel had frustrated Hilbert's program by uncovering an unsuspected limitation in mathematics—true sentences that could not be deduced!

As a young man in the 1920s, Alan Turing took Hilbert's idea of mechanizing mathematics literally. He imagined a machine to do simplified pencil-and-paper calculation on a one-dimensional tape divided into cells, each of which contained just one of a finite set of symbols. A "head," controlled by a few simple rules, moved stepwise from cell to cell, erasing and writing new symbols. Each rule indicated, for each symbol encountered, what symbol to write in its place, which direction to move the head, whether to stop, or else which rule to follow next. Turing demonstrated that all formal mathematical manipulations could be accomplished by this simple device, which came to be known as a Turing machine. Most significantly, he produced rule sets to make Turing machines interpret and obey parts of their tape as other rule sets, also called programs, allowing these *universal* Turing machines to simulate any other machine or even themselves. Analogous to Gödel, Turing showed that there were self-referential questions a universal Turing machine could not answer, but outsiders sometimes could.

Modern computers are universal Turing machines; their internal memory, bearing both program and data, acts as the tape. The Mathematical Objection states that universal machines are inferior to humans because they can be stumped by Gödel questions that humans can answer. Turing challenged the implicit assumption that human beings had no such limitations. We might, for instance, ask Roger Penrose, an honest and thoughtful human, *"Will Roger truthfully answer this question negatively?"* Roger, realizing that answering "No," would make the correct answer "Yes," while his "Yes," would make the proper answer "No," simply cannot answer truthfully—so the correct answer is obviously "No"! *We* can provide the correct answer, but it stymies Roger, so we must be superior to Roger, in just the way the Mathematical Objection makes us superior to machines!

It should be obvious that these are meaningless victories. Roger, or a machine, could just as well have defeated *us* with the same kind of question. Anybody or anything able to think about its own statements and constrained to tell the truth is vulnerable. In practice, of course, humans lie, joke, make mistakes, or contradict themselves, and thus distract or disqualify themselves from Gödelian paradoxes. The Mathematical Objection paints machines as rigid, consistent logical reasoners unable to take any of these escapes. While a few theorem-proving programs aspire to that level of precision, the vast majority of "expert system" and robot-control programs weigh approximate knowledge and contradictory evidence, and they are quite capable of making and tolerating errors. Some have supervisory processes to break them out of contradiction-induced loops.

Mathematical Objectors may answer that underneath every approximate reasoning program lies a rigid, deterministic computer vulnerable to paradox. This fact is probably irrelevant for existing reasoning programs because they are too stupid! A Gödel paradox requires the objects being reasoned about to be capable of expressing the reasoning process itself. Systems not expressive enough are immune. For instance, Euclid's points and lines, lengths and angles are too limited to encode geometric logic, so, unlike arithmetic, geometry has no Gödel sentence. Existing approximate reasoning programs probably escape by having too little time, memory, or functionality to ponder machine-language descriptions of their internals. A complexity limitation may similarly protect human beings from Gödel paradoxes targeted at the neuronal machinery underlying their reasoning. But, perhaps, a neuronal Gödel sentence *could* be streamlined into something verbally expressible. Maybe the world's deepest insight, loveliest vision, catchiest tune, funniest joke, or wickedest spell would obsess a mind, expel all else, and induce a Gödelian seizure!

Like Sleeping Beauty, awakened by a kiss, humans can be snapped out of a reverie by an external disturbance, a crucial point. Gödel's deductive system and Turing's machine are idealizations that uninterruptably follow rigid tracks laid out by a tiny initial definition—no wonder they have limitations. By contrast, humans, robots, and even programs undergoing Turing's conversational test are blessed with unending flows of external information that nudge a steering wheel, causing them to jump the tracks and wander unpredictably over the landscape!

Consider the case of "uncomputable" numbers. A universal Turing machine can simulate any other computer. At the start of a computation, its tape is set up with a finite string of initial program and data, the rest remains blank. As it runs, the machine fills the tape with arbitrarily many symbols, perhaps going on endlessly.

The initial setup is a finite string that can be read as an integer. The ultimate state of the tape can also be interpreted as a number, but because it may have an endless number of digits, it would be a real number like a decimal fraction. One initial setup may cause .33333333 . . . ultimately to be written on the tape, another may write 3.141592653589793 . . . and the rest of the digits of pi. Can every decimal fraction be generated this way? A century ago Georg Cantor, presaging Gödel, surprised the mathematical world by proving that endless decimal fractions are infinitely too numerous to be paired in any way with finite-length integers. But the Turing machine converts each initial setup "integer" into an ultimate "infinite fraction." Only an infinitesimal portion of the decimals can be paired thusly, the rest—essentially all—fractions are "uncomputable." Uncomputable numbers, like unprovable truths, are a fundamental limitation of any finite, closed, deductive system.

Yet, a computer given access to a true randomizing device can generate uncomputable numbers simply by writing an endless stream of random digits in successive cells of its tape. It can also generate random "axioms" not otherwise derivable, which it can "believe" as long as they are not proven inconsistent with other axioms. Random information unlocks the Gödelian cage. The physical world is an even better source of inspiration, because it provides not only endless fresh information but direction. Darwinian evolution depends on noise from the environment to occasionally scramble genomes, and on environmental selection to cull bad changes from the good. Learning similarly depends on variety and consistency in the external world. Both shaped our bodies and our minds and are now helping us shape our machines.

The physical world is not only the unending input, but the inner substance of humans and machines. Rational thought is an abstraction of neuronal chemistry or electronic switching, but chemistry and electronics are abstractions of dancing fields and particles. Physical theories that define this dance can be described by mathematical axioms, and so should be vulnerable to Gödel's paradoxes. From physical descriptions of our mental processes, it should be possible to derive Gödel sentences and Turing questions that are *physically* unanswerable. Although a naive decomposition of our mental operation into basic mathematical physics would be preposterously difficult, there must be mappings of physical events to logical sentences that allow almost any physically unanswerable act (e.g., a skull-crushing blow) to be interpreted as a clever Gödel sentence! Millions of years of bloody experience unambiguously indicate that animals and humans are just as vulnerable as machines to this kind of Mathematical Objection!

4 — *The Argument from Consciousness*

Religious thinkers had the doctrine of soul to separate human from machine, but even the most secular could invoke subjective experience. At the vanguard was a Manchester neurosurgeon, Sir Geoffrey Jefferson, whose 1949 Lister Oration on *The Mind of Mechanical Man* included the following flowery passage:

> Not until a machine can write a sonnet or compose a concerto because of thoughts and emotions felt, and not by the chance fall of symbols, could we agree that machine equals brain—that is, not only write it but know that it had written it. No mechanism could feel (and not merely artificially signal, an easy contrivance) pleasure at its successes, grief when its valves fuse, be warmed by flattery, be made miserable by its mistakes, be charmed by sex, be angry or miserable when it cannot get what it wants.

Turing noted that although we may each know ourselves to be motivated by thoughts and feelings, we have no direct evidence about anyone else. Perhaps all others in the world are actors simply reading lines, or movie projections, mere patterns of light and sound. There is a name for this philosophical position—solipsism—but it has few followers. We usually believe others when they describe motivations and emotions similar to our own. Turing suggested that future machines that behave intelligently and tell us how they feel and why they act will also become accepted as conscious beings.

The next chapter offers a stepwise future evolution of "universal robots" that can report inner feelings. The first generation has motor-perceptual reflexes, the second adds a conditioned learning mechanism, the third a world modeler, and the fourth a general reasoning module. The reflexes give basic physical competence. Internal positive and negative conditioning modules impart a core personality. The modeler of physical and social events, guided by the conditioning system's judgment of possible outcomes, provides foresight and empathy. The reasoner derives abstractions, generalizations, and commentary from the modeler's scenarios.

The robot's psychological-social models are formulated in terms of intentions and feelings of the humans, robots, and animals with which the robot interacts. A simulated situation where the robot's act of locking a door blocks legitimate entry by a human will be negatively reinforced because the modeler indicates that a thwarted human is probably unhappy, triggering a "human unhappiness" conditioning module. Asked why it left the door unlocked, the robot might answer "Roger is outside without a key and would not like to be locked out." Asked why that mattered, the robot could apply its psychological models to itself, note its negative reaction to unhappy humans, and answer, "I don't like to make Roger unhappy."

Even simple robots act on beliefs about the world: whether a path is clear, if a needed object is at a certain location, whether a manipulator is holding something. Correct beliefs result in sensible behavior; incorrect beliefs can cause peculiar actions, like running into walls or grasping thin air. The social models of third- and fourth-generation robots represent beliefs about intentions and feelings. When a robot analyzes its own behavior using these models, it creates beliefs about its own feelings, and may then describe its actions in terms like "I avoid the loading dock because I'm afraid of ledges." Does it then have genuine feelings, does it only believe it has them, or is it only behaving as if it believes it has them? We might ask the same questions of ourselves. In Chapter 7 I argue that feelings, beliefs and thoughts are arbitrary interpretations of physical events. In one accessible interpretation, the robot is simply a mindless mechanical device, behaving as its internal linkages dictate. In another, it has thoughts and beliefs and feelings, and uses those to interpret itself as having them. Turing's argument is that the second interpretation will be adopted naturally by most people to robots that are properly communicative, regardless of how they internally achieve the behavior.

5 — *Arguments from Various Disabilities*

"Professor Turing, I grant machines may do all the things you mentioned, but one can never be constructed to (select one) • be kind • be resourceful • be beautiful • be friendly • have initiative • have a sense of humor • tell right from wrong • make mistakes • fall in love • enjoy strawberries and cream • make someone fall in love with it • learn from experience • use words properly • be the subject of its own thought • have as much diversity of behavior as a man • do something really new." An understandable generalization, Turing surmised, from encounters with thousands of machines, all ugly, narrowly specialized, simple, inflexible, predictable, and obviously mindless.

Before 1950 a straightforward description of what computers were already doing was likely to elicit incredulity, so different were they from the stereotype. Fifty years later, computers are everyday items and intuitive notions of machine nature are being radically altered. As memory, speed, programming, and databases for learning programs have grown, many "nevers" have become commonplace, and others are on the horizon. Although computers are still far less complex than human beings, they are developing so rapidly that confident pronouncements about what machines will never do are rare, and those that are heard have a desperate, whistling-in-the-dark quality.

6 — *Lady Lovelace's Objection*

Charles Babbage, in Victorian England, was the first to conceive an automatic digital computer. His "Analytical Engine" was to be a locomotive-sized structure containing tens of thousands of ten-toothed gears, powered by a steam engine, controlled in detail by pin drums, like a music box, and in major steps by a sequence of program-bearing punched cards, like a Jacquard loom. His ambitious plans caught the interest of Lord Byron's daughter, Ada Lovelace. In 1842, in a detailed description of the machine and its programming, she wrote "The Analytical Engine has no pretensions to *originate* anything. It can do *whatever we know how to order it* to perform."

When fear of "giant brains" jeopardized computer sales in the

1950s, IBM's marketing department promoted the slogan "Computers can do only what their programs specify." They did such a good job, the phrase became a truism by the 1960s.

Lady Lovelace, the first programmer, never had a working computer to trouble her programs. Modern programmers know better. Almost every new program misbehaves badly until it is laboriously debugged, and it is never fully tamed. Information ecologies like time-sharing systems and networks are even more prone to wild behavior, sparked by unanticipated interactions, inputs, and attacks.

Even perfect programs do the unexpected. Mathematical programs produce solutions that their programmers could not have achieved in many lifetimes. Chess programs search branching trees of alternative games to pluck out moves that baffle their programmers and impress chess grandmasters. Programs learning to recognize handwriting and speech process tens of thousands of examples to produce deviously subtle statistical discriminators. Programs have grown from hundreds of instructions in the early days of computing to sometimes millions today. Human programmers, no smarter now than in Babbage's or Turing's time, have had to work at increasing levels of abstraction, entrusting ever larger details to computer search and preprogrammed expertise. Automatic learning will soon open machinery's heart wide to external unpredictability, ending any illusion that computers, and the robots they control, are the mere extensions or tools of human programmers.

7 — The Argument from Continuity in the Nervous System

Macroscopic physical quantities, like the ion potentials, neurotransmitter concentrations, and discharge timings in the nervous system, vary continuously for all practical purposes. Digital computers, by design, mask physical continuity and represent every quantity as a discrete number—they count rather than measure. Many people feel that counting is less powerful than measuring, if more rigidly precise, and so computers are inherently more limited than nervous systems.

Decades before information was transmitted, stored and processed digitally, it moved in continuous form on wire, radio, records, film, tape, and "analog" computers. Communications engineers, who studied analog techniques as they now study digital, found a fundamental imprecision. Few analog processes are repeatable beyond the fourth decimal place because they are perpetually jostled by the microscopic motions of heat. Digital machinery, on the other hand, precisely handles ten digits of precision when it allocates thirty switches to a quantity. When machinery was handmade, it was easier to build and adjust one big, slow analog knob than thirty small, fast digital switches, but today's photographic integrated-circuit techniques make repetition of circuit elements easy. Precision, mathematical agility, and errorless storage and transmission make signals far more potent in digital than original analog form. These merits, and rapidly declining cost, have assured digital's ascendancy in computers, communications, audio, and video. Analog continues to be used in places where it retains an advantage—for instance, in robot end-effectors and sensors.

The naive impression that analog is superior to digital probably arises from human awkwardness with arithmetic compared to our nimbleness in perceiving positions, sizes, and other quantities. Turing noted that digital computers could easily simulate human imprecision, to fool judges in his test, by deliberately perturbing digital quantities with small random offsets.

8 — The Argument from Informality of Behavior

"Professor Turing, human beings encounter many unexpected circumstances in their lives and respond to them in novel, unpredictable ways. No set of rules programmed into a machine could prepare it for every eventuality. Therefore, humans could not possibly be machines, and machines could not behave like humans." Turing found this argument hard to fathom, but he supposed it might result from an ambiguous use of the word "rules." At one extreme there are social rules, such as "Don't cross the street on a red light"; at the other, there are natural laws, such as "A mass accelerates in the direction of an applied force." Even Emily Post could not compile a list of social rules covering every eventuality, but we expect natural law to be all-inclusive.

A machine is bound to react in *some* way to any circumstance, foreseen or unforeseen. A robot that boringly repeats a simple task whenever its "Go" button is pressed will tumble and smash excitingly when pushed off a cliff. The most interesting behavior happens between the extremes of order and chaos, when responsive mechanisms dance with complicated environments. Wire a toy robot so that a touch on the left causes a veer to the right, and touch on the right a veer to the left, and the machine will meander engagingly, redirected by each obstacle. Give it similar connections to optical sensors and it will cavort mysteriously, avoiding obstacles and chasing lights. A more complex robot might have a spectrum of responses to inputs, from simple reflexes to elaborate preplanned action procedures. In a rich environment its behavior would intertwine spontaneity with rule-following, just like a human being.

9 — The Argument from Extrasensory Perception

Turing's ninth argument seems odd today, when scientists dismiss prophets, witches, magicians, fortune-tellers, and psychics as charlatans. The situation was different when Turing wrote. In 1935 Joseph Rhine founded a "parapsychology" laboratory at Duke University in North Carolina to conduct the first controlled scientific experiments on the human ability to sense and influence remote or future events by thought alone. Rhine's laboratory, and its imitators, reported evidence for these "paranormal" abilities using methodologies that convinced a significant portion of the scientific community. By 1950 the paranormal was a frequent topic in informal scientific discussions, including those about intelligent machines.

In one Duke experiment, sequestered subjects guessed the identity of cards viewed by someone elsewhere. Some did better than chance, perhaps by reading the viewer's mind. Since deterministic computers would fail the imitation test if pitted against a mind-reading human, Turing suggested giving the machine a physical randomizer that might be psychically influenced by the judge, or tightening up the test by putting the competitors in "telepathy-proof" rooms!

Rhine's glow faded in subsequent decades, when careful repetitions by others failed to duplicate his results and third-party analyses of the original experiments uncovered subtle biases—for instance outcomes selectively not reported because promising subjects were "not ready" at the beginning of a session or "weary" at the end. Several major experiments since Rhine's have claimed evidence for the paranormal, only to succumb themselves to critical examination. One, conducted between 1983 and 1989 by the psychologists Daryl Bem and Charles Honorton of Cornell and Edinburgh Universities, still stands.

Credulity, psychological quirks, social role playing, subliminal cueing, probability misjudgments, experimental errors, and charlatanry alone may explain worldwide belief in the supernatural, but not yet conclusively. The study of life and mind is in such early days that paranormal effects could be partying unsuspected under science's nose. Biological evolution undogmatically exploits any source of useful information. Microscopic magnetite crystals orient some bacteria and birds. Pressure, humidity, low-frequency sounds, and pheromones are known to affect human mood subconsciously. Chapter 6 speculates that time travel could be common, yet elusive to paradox-threatening direct examination, perhaps manifesting itself as coincidences between apparently unconnected events. A nervous system could distill information from mysterious coincidences within itself as easily as from subtle sensory patterns. An ideal digital computer, its parts operating in rigid lockstep, would not be open to such effects, unless they are present in the initial input. On the other hand, a real machine with sensors like cameras, microphones, and radio receivers that register both signal and noise probably would. Maybe an advanced psychic sense-organ for robots would have an array of sensitive detectors tightly shielded from outside influence, reporting to a powerful learning program that statistically correlates their hiss of chance with the robot's life experiences. Or maybe not.

Consciousness Raising

Meet Roger, a nice fellow, but a biological chauvinist

Roger believes his Dog has conscious feelings

But he thinks his house robot is just a dumb machine

With no feelings at all

One day a package arrives in the mail

It has something for everyone in the house.

Sometimes Roger lives dangerously

The software makes the machine care about what Roger thinks about it. It recalls cruel words spoken in past. Then they didn't matter, but now they induce action

Please, Roger, it bothers me that you don't think of me as a real person. What can I do to convince you? I am aware of you, and I am aware of myself. And I tell you, your rejection is almost unbearable

Roger, soft-hearted, relents

4

Universal Robots

Affordable computers raced past 1 MIPS in the 1990s, but the benefits have been slow to reach commercial robots. There are hundreds of millions of computers, but only a few hundred thousand robots, many over a decade old. The perceived potential of robotics is limited, and the engineering investment it receives consequently modest. Advanced industrial robots are controlled by last generation's 1- to 10-MIPS computers, less intelligent than an insect. High-speed robot arms use the processing power to precisely plan, measure, and adjust joint motions about one hundred times per second. Mobile robot computers are kept busy tracking a few special navigational features, calculating the robot's position, checking for obstacles and planning and adjusting travel about ten times per second. As with insects, success depends on special properties of the environment: precise location and timing of workspace components for factory arms, correct placement and programming of the navigational features, and absence of too much obscuring clutter for mobile machines. Insects behave more interestingly than present robots because evolutionary competition has drawn them into wickedly risky behavior. Insect survival relies as much on mass reproduction as on individual success; only a tiny, lucky fraction of each generation lives long enough to produce offspring. Even so, similar limitations surface in both domains. A single misplaced beacon draws a Stanford Hospital robot down a stairwell, and a moth spirals to its death, navigation system fixated on a streetlight rather than the moon. To reduce such risks, $100,000 machines are designed to be plodding dullards, restricted to carefully mapped and marked surroundings.

Robot systems are now installed, debugged, and updated by trained specialists, who measure and prepare the workspace and tailor job- and site-specific control programs. Few jobs are large and static enough to warrant such time-consuming and expensive preparations. If mobile robots for delivery, cleaning, and inspection could be unpacked anywhere and simply trained by leading them once through their task, they would find thousands of times as many buyers. The performance of this decade's experimental robots strongly suggests that 10 MIPS is inadequate to do this sufficiently well. Insectlike 100-MIPS machines, with programs that build coarse 2D maps of their surroundings, or track hundreds of points in 3D, might, just possibly, free-navigate tolerably in some circumstances. One thousand MIPS computers, mental matches for the tiniest lizards, can manage multi-million cell 3D grid maps and probably *are* adequate to guide free-roving robots. The basic idea would be to align current maps to old maps stored during training: the more content in the maps, the less chance of error. One thousand MIPS is just about to appear in personal computers, and thus in research robots. It may show up in commercial robots early in the next decade, multiplying their numbers along with their usefulness.

Utility robots for the home could then follow, maybe around 2005, drawn by an assured mass market for affordable machines that could effectively keep a house clean. Early entrants will probably be very specialized, like the conceptual autonomous robot vacuum cleaner in the following illustration. This design senses its world with a triangle of tiny video cameras on each of four sides, is controlled by a 1,000-MIPS computer, and moves in any direction (and scrubs) using three individually steered and driven wheels. About once a second, it stereoscopically measures the range of several thousand points in its vicinity and merges them into a three-dimensional "evidence grid" map, as described in Chapter 2. The grid gives the robot a spatial intelligence comparable to a tiny lizard's, but more precise. Taken out of its box and activated in a new home, the robot memorizes its surroundings in 3D. Then, perhaps, it asks "when and how often should I clean this room, and what about the one beyond the door?" Its spatial comprehension would keep it doing its job and out of trouble for years at a stretch.

Perhaps it maintains a "web page" by wireless connection, where its maps and schedules can be examined and altered.

Universality

Commercially successful robots will engender a growing industry and more capable successors. The simple vacuum cleaner may be followed by larger utility robots with dusting arms. Arms may become stronger and more sensitive, to clean other surfaces. Mobile robots with dexterous arms, vision and touch sensors, and several thousand MIPS of processing will be able to do various tasks. With proper programming and minor accessories, such machines could pick up clutter, retrieve and deliver articles, take inventory, guard homes, open doors, mow lawns, or play games. New applications will spur further advancements when existing robots fall short in acuity, precision, strength, reach, dexterity, or processing power. Capability, numbers sold, engineering and manufacturing quality, and cost effectiveness will increase in a mutually reinforcing spiral.

Rossum's Universal Robots was the title of an influential internationally performed play written in 1921 by a Czech playwright, Karel Čapek. At his brother's suggestion, Čapek coined the term "robot" from the Czech word for hard, menial labor. In the play, a universal robot was an artificial human being built to do drudge work of any sort, especially in factories.

In 1935 Alan Turing instantiated David Hilbert's concept of mechanizing mathematics. He designed conceptual machines that moved along a tape, reading and writing discrete symbols according to simple fixed rules, and showed they could do any finite mathematical operation. He also demonstrated that there exist particular machines that interpret the initial contents of their tape as a description of any other machine and proceed, slowly, to simulate this other machine. Such *Universal Turing Machines* inspired some of the first electronic computers and still serve as mathematical tools for studying computation.

Computers are the physical realization of Turing's universal machines, but they are universal only in symbol manipulation—paperwork. A universal *robot* extends the idea into the realm of physical perception and action. Because there is a far greater variety

Dustbot

This conceptual design for an automatic home vacuum cleaner of the near future is intended to function with very little instruction from its owner. It has omni-directional wheels, stereoscopic eyes on all faces, and 1,000 MIPS of processing programmed to give it a 3D sense of space. The robot is shown vacuuming under furniture, at a docking station regurgitating accumulated dust and recharging with retracted nozzle, moving at an angle to clean room edges and corners, and topping up its batteries at a handy socket, using an optional "field recharging arm."

and quantity of physical work to do in the world than paperwork, universal robots are likely to become far more numerous than plain universal computers as soon as their capabilities and costs warrant. Advanced utility robots reprogrammed for various tasks will be a partial step in that direction. Call them the accidental, zeroth generation of universal robot. Subsequent generations will be designed from the start for universality.

First-Generation Universal Robots

Estimated time of arrival: 2010

Processing power: 3,000 MIPS (lizard-scale)

Distinguishing feature: General-purpose perception, manipulation, and mobility

A robot's activities are assembled from its fundamental perception and action repertoire. First-generation robots will exist in a world built for humans, and that repertoire most usefully would resemble a human's. The general size, shape, and strength of the machine should be humanlike, to allow passage through and reach into the same spaces. Its mobility should be efficient on flat ground, where most tasks will happen, but it should somehow be able to traverse stairs and rough ground, lest the robot be trapped on single-floor "islands." It should be able to manipulate most everyday objects, and to find them in the nearby world.

Two-, four-, and six-legged robots of all sizes, able to cross many types of terrain, are becoming common in research. Many, however, are powered externally via wires, and those that carry their own power move slowly and only for short distances on internal power. The mechanics are complex and heavy, with many parts and linkages and at least three motors per leg. Developments in materials, design, motors, power sources, and automated manufacture may ultimately change the economics, but, for the immediate future, legged locomotion cannot compete in performance, cost, or reliability with wheeled vehicles in mostly flat work areas. Typically, a wheeled robot can travel for a day on a battery charge, while a walker, going more slowly, expires in an hour. Yet simple wheels cannot negotiate stairs—a decades-old dilemma for inventors of powered wheelchairs. Among the patented stair-climbing solutions are pivoted tracks, three-spoked wheels ending in smaller wheels, and specialized mechanical feet of many varieties. All have significant costs in weight, efficiency, or maneuverability over plain wheeled vehicles. Robots needed as "frequent climbers" may be configured with such mechanics, but the majority probably will simply roll.

Elevators and ramps mandated for wheelchairs will also provide access to nonclimbing robots. Where such conveniences are lacking, universal robots may be able to lay their own ramps, temporarily dock with stair-climbing mechanisms, or even winch themselves up and down on cables. They have the advantage over humans of being exactly as patient as their control programs specify.

Robots that simply travel have uses, but many jobs call for grasping, transporting, and rearranging ingredients, parts, and tools. Fixed industrial arms, with about six rotary or sliding joints, have good reach and agility but are too large and heavy for mobile robots. NASA, the U.S. space agency, has supported research on lightweight robot arms made of materials like graphite composites, with compact high-torque motors and controls that compensate for flexing in the thin limbs. Cheaper derivatives of these designs could make excellent arms for universal robots. Since many jobs require bringing pairs of objects into contact, robots will probably be able to have several arms, of various sizes.

Human hands are much more complicated than arms, and harder to imitate in robots. Most industrial robot grippers work like miniature vices, others use fixtures shaped for specific jobs, and a few change their "end effectors" from task to task. Inventors worldwide have devised many clever hands that mold themselves to odd objects, but lacking the ability to control their detailed forces, they fail in some grips, and none are used in practice. More elaborate hands, humanlike and otherwise, with multiple individually controlled fingers, have been demonstrated by robotics researchers. With their many segments, linkages, and motors, they tend to be heavy and expensive, and controlling them to sensitively grasp, grip, orient, and fit objects is a difficult area of ongoing research. Because they operate only occasionally, and exert modest forces over short distances, hands can afford to be less energy efficient than mobility systems or arms and so can use types of "muscle" chosen to be more compact than efficient. Particularly compact are "shape-memory" alloys, metals which, easily bent at room temperature, return with great force to their original shape when heated. Given the difficulties, the first generations of universal robot will probably make do with simple, imprecise hands, leaving fine dexterity for the future.

The navigational needs of universal and housecleaning robots are similar. A robot will perceive its surroundings with sensors, probably video cameras configured for stereoscopic vision, and construct a 3D mental map. From the map it will recognize locations, plan trajectories, and detect objects by shape, color, and location. This latter ability will be more important, and thus better developed, in universal robots than in housecleaners. Universal robot maps may be higher resolution, giving them a sharper sense of general spatial awareness. They will probably need an especially fine map of the volume around their hands to precisely locate work objects and visually monitor manipulation. An interesting technique, used successfully in research projects and becoming more practical as cameras shrink, is to put video cameras on the hands themselves, in the palms or on the fingers.

A few thousand MIPS is just enough computing power for a moving robot to maintain a coarse map of its surroundings and use it for locating itself relative to trained itineraries and to plan and control driving. When not traveling, there is power enough to construct a fine map of a manipulator workspace, to locate particular objects, and to plan and control arm motions. Speech and text recognition is advanced enough today that robots of 2010 will surely be able to converse and read. They will also be linked with the Internet, giving them a kind of telepathy. Using it, they will be able to report and take instructions remotely and to download new programs for new kinds of work. Information security will be even more important then than now, because carelessly or maliciously programmed robots will be physically dangerous.

Universal robots may find their very first uses in factories, warehouses, and offices, where they will be more versatile than the older generation of robots they replace. Because of their breadth of applicability, their numbers should grow rapidly and their costs decline. When they become cheap enough for households, they will extend the utility of personal computers from a few tasks in the data world to many in the physical world. Perhaps a program for basic housecleaning will be included with each robot, as word processors were shipped with early personal computers.

As with computers, some applications of robots will surprise their manufacturers. Robot programs may be developed to do light mechanical work (e.g., assembling other robots,), deliver warehoused inventories, prepare specific gourmet meals, tune up certain types of car, hook patterned rugs, weed lawns, run races, play games, arrange earth, stone, and brick, or sculpt. Some tasks will need specialized hardware attachments like tools and chemical sensors. Each application will require its own original software, very complex by today's computer program standards. The programs will contain modules for recognizing, grasping, manipulating, transporting, and assembling particular items, perhaps modules developed via learning programs on supercomputers. In time, a growing library of subtask modules may ease the construction of new programs.

A first-generation robot will have the brain power of a reptile, but most application programs will be so hard pressed to accomplish their primary functions that they will endow the robot with the personality of a washing machine.

Second-Generation Universal Robots

Estimated time of arrival: 2020

Processing power: 100,000 MIPS (mouse-scale)

Distinguishing feature: Accommodation learning

First-generation robots will be rigid slaves to inflexible programs, relentless in pursuing their tasks—or repeating their errors. Their programs will contain the frozen results of learning done on bigger computers under human supervision. Except for specialized episodes like recording a new cleaning route or the location of work objects, they will be incapable of learning new skills or adapting to unanticipated circumstances. Even modest alterations of behavior

P2, an embryonic universal robot

Not a man in a spacesuit, but a self-contained research robot developed over a decade by a group of thirty engineers at Honda Motors of Japan, perhaps as a hedge against future reverses in the automobile market. The "backpack" contains power and computing. The machine has fully functional arms and camera eyes, and can find stairs and move objects. Its most advanced skill, so far, is walking, on flat and sloped ground, and up and down stairs. It is as humanlike in its motion as in its appearance. If pushed, it shifts its posture or begins to walk to keep equilibrium. It stands 180 cm tall and weighs 210 kilograms. At several million dollars, with a 15-minute battery lifetime, it is too expensive and power-consumptive to be practical, but its development continues and it is surely a precursor of future universal robots. A slightly smaller successor, the P3, with a 25-minute battery life, is probably the most advanced self-contained robot today.

will require new programming, probably from the original software suppliers.

Second-generation robots, with thirty times the processing power, will learn some of their tricks on the job. Their big advantage is adaptive learning, which "closes the loop" on behavior. Each robot action is repeatedly adjusted in response to measurements of the action's past effectiveness. In the simplest technique, programs will be provided with alternative ways, both in the large and the small, to accomplish steps in a task. Alternatives that succeed become more likely to be invoked in similar circumstances, while alternatives that fail become less probable. "Statistical learning" is another approach, in which a large number of behavior-modifying parameters (e.g., weights in simulated neural nets) are repeatedly tweaked to nudge actual behavior closer to an ideal. Programs for second-generation robots will use many such learning techniques, creating new abilities—and new pitfalls.

Some programs may learn with human assistance. To teach a robot to recognize shoes, its owner might assemble some shoes in a cluttered room, and point them out. The robot, running an object-learning program, will note the shape and color of both the shoes, and of other objects, to train a statistical classifier to distinguish shoes from non-shoes. If the robot later meets a still-ambiguous object, say a gravy boat, it may ask the owner's opinion, to further tune the classifier. Similar programs might learn minor motor skills by first recording movements as the robot is lead by hand through the motions, then optimizing the details in practice playbacks. Learned modules, though tedious to create, will probably be easy to package and employ. A general tidying-up program, for instance, could be modified to gather shoes by grafting in a shoe-recognizing module.

Second-generation robots will occasionally be trained by humans, but more often will simply learn from their experiences. The behavior loop is closed in the latter case by a collection of constantly running *conditioning* programs, or modules, that watch out for generally desirable or undesirable situations and generate signals that act on the task-oriented programs that actually control the robot. Each major and minor step in a second-generation application program will have a variety of alternatives: a grasp can be underhand or overhand, harder or softer, an arm movement can be faster or slower, lifting can be done with one arm or two, an object can be sought with one of several recognition modules, two objects can be assembled in a number of locations, and so on. Each alternative will have associated with it a relative probability of being chosen. Whenever a conditioning module issues a signal, the probabilities of recently executed alternatives are altered. Conditioning signals come in two broad categories: positive, which raise the probabilities, and negative, which lower them. What situations the conditioning modules respond to, and what type and strength of signal they send, is entirely up to the robot's programmers. Some choices seem sensible. Strong negative signals should result from collisions, from objects dropped or crushed, when a task fails or when the robot's batteries are nearly discharged. A near miss, a task completed unusually slowly, or an excessive acceleration might trigger weak negative signals. A rapidly completed task, charged batteries, or an absence of negative signals could produce positive signals. Some conditioning modules may be tied to the robot's speech recognition, generating positive signals for words of praise and negative ones for criticism. Speech recognizers in the 1990s struggle merely to identify the words being spoken. By 2010 they should be able to identify speakers and emotional overtones as well. A second-generation robot with a "general-behavior" program having options for almost any action at every step could probably be slowly trained for new tasks purely through Skinnerian conditioning, like a circus bear. More practical "user-trainable" programs will probably allow the key steps to be directly specified by voice instruction and leading by hand, with the conditioning saved for fine-tuning.

If a first-generation robot working in your kitchen runs into trouble, say, fumbling a key step because a portion of the workspace is awkwardly small, you have the option of abandoning the task, changing its environment, or somehow obtaining altered software that accomplishes the problematic step in a different way. A second-generation robot will make a number of false starts, but most probably will find its own solution. It will adjust to its home in thousands of subtle ways and gradually improve its performance. While a first-generation robot's personality is determined entirely by the sequence of operations in the application program it runs at the moment, a second-generation robot's character is more a product of the suite of conditioning modules it hosts. The conditioning system might, in time, censor an entire application program, if it gave consistently negative results.

Learning is dangerous because it leaps from a few experiences to general rules of behavior. If the experiences happen to be untypical, or the conditioning system misidentifies the relevant conditions, future behavior may be permanently warped. An important element of the ALVINN automatic driving system of Chapter 2, which learns to follow a particular type of road by imitating human driving, is a carefully contrived augmentation of the actual camera scenes. Dean Pomerleau's program transforms each real training image and its associated steering command into thirty different ones by geometrically altering them in a way that approximates shifting the truck up to a half lane-width in either direction. Without this stratagem, learned steering is dangerously erratic away from the center of the lane. The program adds other artifacts to teach the system to ignore areas beyond the road edge and distant traffic. Other flexible learning programs, including those in my own research, behave similarly: they are prone to learn dangerous irrelevancies and miss important features if their training experiences are not properly structured. Occasionally, as in the transition from ALVINN to RALPH, enough is known in advance to hardwire in reasonable restraints. More often, though, the necessary information is unavailable at programming time.

Even animals and humans, veterans of tens of millions of years of evolutionary honing, are vulnerable to inappropriate learning. A friend's dog, struck by a car while crossing a road, continued wantonly to cross roads. It would, however, under all circumstances, avoid the region of sidewalk from which it had set out before the accident. People and animals can be victims of self-reinforcing phobias or obsessions, instilled by a few traumatic experiences or abusive treatment in formative years. A first-generation robot, "trained" off-line under careful factory supervision, will suffer no permanent ill effects from events that interfere with its work but leave it physically undamaged. A second-generation robot, though, could be badly warped by accidents or practical jokes. It might remain impaired until its conditioned memory is cleared, restoring it to naive infancy. Or perhaps, sometimes, robot psychologists could slowly undo the damage. Devising conditioning suites that teach quickly, but resist aberrations, will be among the greatest challenges for programmers of second-generation robots. There is probably no perfect solution, but there are ways to approach the problem.

Second-generation robots of 2020 will have onboard computers as powerful as the supercomputers that learned for first-generation machines a decade before. But by 2020 supercomputers will be proportionally more powerful and will themselves play a background role. The many individual programs of a conditioning suite, each responding to some specific stimulus, interact with one another and with the robot's control programs and environment in ways that will be far too entangled to anticipate accurately. It would be possible to evaluate particular suites by trying them out in robots, the acid test in any case, but that would be a slow and dangerous way to sift a large number of rough candidates. Some would certainly behave in unexpected ways that could damage the robot or even endanger the testers.

Faster and safer initial screenings might be done in factory supercomputer simulations of robots in action. To be of value, simulations would have to be good models, predicting accurately such things as the probability that a given grip can lift a particular object, or that a vision module can find a given something in a particular clutter. Simulating the everyday world in full physical detail will still be beyond computer capacity in 2020, but it should be possible to approximate the results by generalizing data collected from actual robots, essentially to learn from the working experience of real robots how everyday things behave. A large systematic collection effort under human supervision will probably be necessary lest there be too many gaps or distortions. A proper simulator would contain at least thousands of learned models for various basic interactions (call them *interaction models*), in what amounts to a robotic version of common-sense physics. It would be a perceptual/motor counterpart to the purely verbal framework being collected by the Cyc effort to endow reasoning programs with common sense, mentioned in Chapter 2.

The simulators could be used to automatically find effective conditioning suites, in effect, to learn how to learn. A suite might be evaluated on a simulated robot running a few favorite application programs in a simulated household for a few simulated days. Repeatedly, suites that produced particularly effective and safe work would be saved, modified slightly, and tried again, while those that did poorly would be discarded. This kind of process, called a genetic algorithm, is a computerized version of Darwinian evolution. It is sometimes the most effective way to optimize when the relation between the adjustments (the choice and settings of conditioning modules, in this case) and the quantity to be optimized (robot performance) has no simple model.

Third-Generation Universal Robots

Estimated time of arrival: 2030

Processing power: 3,000,000 MIPS (monkey-scale)

Distinguishing feature: World modeling

Adaptive second-generation robots will find jobs everywhere and may become the largest industry on earth. But teaching them new skills, whether by writing programs or through training, will be very tedious. A third generation of universal robots will have onboard computers as powerful as the supercomputers that optimized second-generation robots. They will learn much faster because they do much trial and error in fast simulation rather than slow and dangerous physicality. Once again, a process done by human-supervised supercomputers at the factory in one robot generation will be improved and installed directly on the next generation, and once again new opportunities and new problems will arise.

With a fast-enough simulator, it will be possible for a robot to maintain a running account of the actual events going on around it—to simulate its world in real time. Doing so requires that almost everything the robot senses be recognized for the kind of object it is, so that the proper interaction models can be called up. Recognizing arbitrary objects by sight is as difficult as knowing how they will interact. It will require modules specially trained for each kind of thing (call them *perception models*). Some perception models may already have been developed for second-generation factory simulators to help automate the tedious job of creating simulations of

A conceptual universal robot

An omnidirectional wheelbase allows flexible movement on flat floors. Elevators, ramps, hoists, or special carrier carts would be used to change floors, climb stairs, or traverse very rough ground. The central post, on a swivel mount, is a "bus" that provides mechanical support, power, and control for a changeable suite of manipulators, sensors, and other accessories that rotate and ride up and down its length. One such accessory, an array of miniature cameras, gives the robot 360 degrees of stereoscopic vision, primarily for navigation. Manipulator-mounted cameras provide precise views of work objects. To reach greater heights, the robot itself can extend its post by attaching new segments. Batteries and computers in the base supply power, control, and stability. Major structural members, like the arms, are made of strong, light, composite materials. Lightweight, high-torque electric motors drive the large motions, like the wheels and arms. Even lighter, though less efficient, actuators like shape-memory metals drive the many motions of the fingers. These innovations combine to give the robot roughly the size, weight, strength, and endurance of a human in a spindly structure that resembles the cartoon broomsticks in Disney's "The Sorcerer's Apprentice."

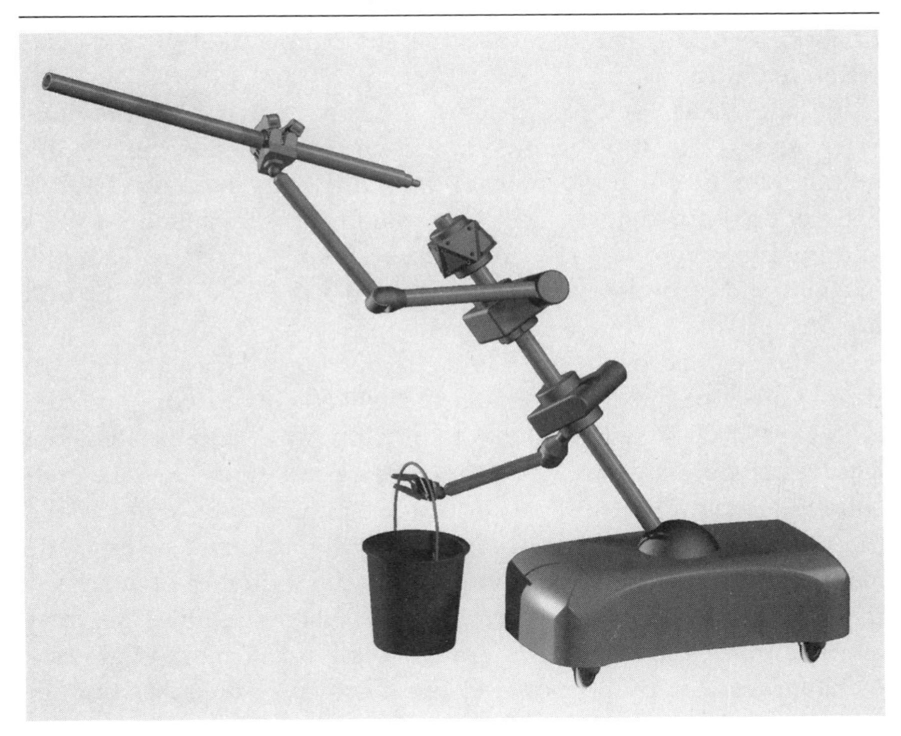

robot workspaces. An additional effort to fill gaps and systematize the factory repertoire will surely be necessary to prepare it for fully automatic use in the third generation. Perception models will allow a robot's three-dimensional map of a room to be transformed into a working model where each object is identified and linked with its proper interaction models.

A continuously updated simulation of self and surroundings gives a robot interesting abilities. By running the simulation slightly faster than real time, the robot can preview what it is about to do, in time to alter its intent if the simulation predicts it will turn out badly—a kind of consciousness. On a larger scale, before undertaking a new task, the robot can simulate it many times, with conditioning system engaged, learning from the simulated experiences as it would from physical ones. Consequently well trained for the task, it would likely succeed the first time it attempted it physically—unlike a second-generation machine, which makes all its mistakes out in real life.

When it has some spare time, the robot can replay previous experiences, and try variations on them, perhaps learning ways to improve future performance. A sufficiently advanced third-generation robot, whose simulation extends to other agents—robots and people—would be able to observe a task being done by someone else and formulate a program for doing the task itself. It could imitate.

A third-generation robot might also be induced to invent its own simple programs in response to a specialized conditioning module whose rewards are proportional to how nearly a sequence of robot actions achieves a desired end. Repeated simulations of a general-behavior program with such a "teacher" might gradually shape the program into one that accomplishes the specified result. The teacher can be quite simple compared to the sequence of actions it induces, and could be constructed via voiced commands. The statement "*Put the glass on the table*" might create a conditioning module whose reward is proportional to the distance of the glass bottom from the tabletop. Steered by this, and by standard modules such as ones that generate negative signals if the glass spills or drops, repeated simulations may devise a sequence of arm motions that does the job.

There are complications. A simulator will be dangerously misleading unless it accurately models objects and events. A newly delivered robot is unlikely to have good representations of the personalized knickknacks it freshly encounters and will need some way of learning, perhaps by assigning to new objects perception and interaction models from a set of generic classes. Then it would tune the interaction models of particular objects whenever a real event and its simulation differed. Since it would be dangerous to start a robot on a complex task before it had good models of the items involved, third-generation robots will require noncritical "play" periods wherein things are handled, spaces explored, and minor activities attempted, simply to tune up the simulation.

Although they will be able to adapt, imitate, and create simple programs of their own, third-generation robots will still rely on externally supplied programs to do complicated jobs. Since their motor and perceptual functions will be quite sophisticated, and their memories and potential skills large, it will be possible to write wonderfully elaborate control programs for them, accomplishing large jobs, with nuances within nuances. It will be increasingly difficult for human programmers to keep track of the many details and interactions. Fortunately, the task can be largely automated. Shakey, the first computer-controlled mobile robot, mentioned in Chapter 2, had at its heart a reasoning program called STRIPS (STanford Research Institute Problem Solver) that expressed the robot's situation and capabilities as sentences of symbolic logic. STRIPS solved for the sequence of actions that achieved a requested result as a proof of a mathematical theorem. On Shakey's 0.3 MIPS computer, neither the theorem prover nor the sensory processing that fed it could handle the complexity of realistic situations, and Shakey was limited to maneuvering around a few blocks.

Despite Shakey's limitations in 1969, the idea of planning robot actions with a theorem prover was sound. Given a correct description of the initial and desired state of the world, and the robot's abilities, along with enough time and space to work, a theorem prover will find an absolutely correct solution, of arbitrary generality, subtlety, and deviousness, if one exists. By the time of the third universal-robot generation, supercomputers will provide 100,000,000 MIPS, and (thanks to continuing progress in

the top-down Artificial Intelligence industry) programs will exist that will be able to perform STRIPS-like reasoning with real-world richness. So factory supercomputers in 2030 will accept complex goals (find a sequence of robot actions that assembles the robot described in the following design database), and compile them via theorem provers into wonderfully intricate control programs for third-generation robots which will, in the field, adapt them to their actual circumstances.

Fourth-Generation Universal Robots

Estimated time of arrival: 2040

Processing power: 100,000,000 MIPS (human-scale)

Distinguishing feature: Reasoning

In the decades while the "bottom-up" evolution of robots is slowly transferring the perceptual and motor faculties of human beings into machinery, the conventional Artificial Intelligence industry will be perfecting the mechanization of reasoning. Since today's programs already match human beings in some areas, those of forty years from now, running on computers a million times as fast as today's, should be quite superhuman. Today's reasoning programs work from small amounts of unambiguous information prepared by human beings. Data from robot sensors such as cameras is much too voluminous and noisy for them to use. But a good robot simulator will contain neatly organized and labeled descriptions of the robot and its world, ready to feed a reasoning program. It could state, for instance, if a knife is on a countertop, or if the robot is holding a cup, or even if a human is angry.

Fourth-generation universal robots will have computers powerful enough to simultaneously simulate the world and reason about the simulation. Like the factory supercomputers of the third generation, fourth-generation robots will be able to devise ultrasophisticated robot programs, for other robots or for themselves. Because of another gift from the Artificial Intelligence industry, they will also be able to understand natural languages. Disembodied language understanders may use a verbal common-sense database similar to the one being developed by the Cyc project, where the meaning of words is defined in reference only to other words. Fourth-generation robots will understand concepts and statements more deeply, through the action of their simulators. When someone tells a robot *"The water is running in the bathtub,"* the robot can update its simulation of the world to include flow into the unseen tub, where a simulated extrapolation would indicate an undesirable overflow later and so motivate the robot to go to turn off the tap. A purely verbal representation might accomplish the same thing if it included the statements such as "A filling bathtub will overflow if its water is not shut off." However, a few general principles in a simulator, interacting in combinations, can substitute for an indefinite number of sentences.

Similarly, a reasoning program, making inferences about physical things, might be enhanced by a simulator. Candidate inferences would be rejected if they failed in a parallel simulation of a typical case, and, conversely, persistent coincidences in the simulation could suggest statements that can be proved or assumed. The robot would be visualizing as it listened, spoke, and reasoned. A modest but very successful version of such an approach was used in one of the earliest Artificial Intelligence programs, a geometry theorem prover by Herbert Gelernter in 1959. Starting with the postulates and rules of inference in Euclid's "Elements," Gelernter's program proved some of the theorems, using algebraic "diagrams" to eliminate false directions in the proofs. Before attempting to prove two triangles congruent in a certain construction, for instance, the program would generate an example of the construction, using random numbers for the unspecified quantities, and measure the resulting triangles. If the specific diagramed triangles did not match within the precision of the arithmetic, the program abandoned the attempt to prove them congruent in general.

Simulator-augmented language understanding and reasoning may be so effective in robots that it will be adopted for use in plain computer programs, "grounding" them in the physical world via the experiences of the robots that tuned the simulators. In time the distinction between robot controllers and disembodied reasoners will diminish, and reasoning programs will sometimes link to robot bodies to interact physically with the world, and robot minds will sometimes retire into large computers to do some intense thinking off-line.

A fourth-generation robot will be able to accept statements of purpose from humans and "compile" them into detailed programs that accomplish the task. With a database about the world at large, the statements could become quite general—things like "earn a living," "make more robots," or "make a smarter robot." In fact, fourth-generation robots will have the general competence of human beings and resemble us in some ways, but in others will be like nothing the world has seen. As they design their own successors, the world will become ever stranger. That is the subject of the next chapter.

The Inner Lives of Robots

The four generations of universal robot controllers resemble four hundred-million-year stages in vertebrate brain evolution, marked roughly by enlargement of the brain stem, the cerebellum, the midbrain, and the cerebrum. The development pace is fast forwarded ten-millionfold. Since the fourth-generation robot resembles us in evolution and behavior, we can ask whether it has an internal mental life anything like ours. Is it conscious of its existence? Does it have emotions?

The conventional answer to these questions is a reflexive *"No!"* Machines, after all, are cold and unfeeling. Further, the idea of machinery with a conscious mental inner life frightens or enrages some people—an understandable visceral reaction, as the concept clashes with a deep primordial view of the nature of things. Similar vehemence greeted predictions of space travel early in the twentieth century. Space travel violated the self-evident dichotomy of the terrestrial and the celestial, a sacred distinction in most religions, whose abrogation, if possible at all, would surely upset the natural order, with horrible consequences. Thoughtful machinery violates the equally obvious and sacred dichotomy of the living and the dead, a difference embedded in our mentality. The skills for interacting with living things, with feelings, memories, and intentions, are utterly different from the techniques for shaping insensitive dead matter.

Although animate machines have existed for several centuries, they have acted more like inanimate things, with no awareness of past or future, responding to the skill but not the character of their handlers. But past machines were simpler than bacteria, and are as misleading a guide to future robot mentality as bacteria are to human psychology. Ancient thinkers theorized that the animating principle that separated the living from the dead was a special kind of substance, a spirit. In the last century biology, mathematics, and related sciences have gathered powerful evidence that the animating principle is not a substance, but a very particular, very complex organization. Such organization was once found only in biological matter, but is now slowly appearing in our most complex machines. In the old metaphor, we are in the process of inspiriting the dead matter around us. It will soon be our honor to welcome some of it to the land of the living, however upsetting that may be to our traditional categories.

External Manifestations

Manifestations of internal life will mount with the robot generations. Good programs for first-generation machines will have responses for likely contingencies. Their robot hosts will perform their job, but also react to obstacles, missing supplies, little accidents, and so forth. Within tightly circumscribed boundaries, they will seem to have genuine awareness of their circumstances. Outside of that range, the impression will collapse, as their program repertoire runs out and they stop with an error message. The impression will also fade over time, as their responses are seen never to vary.

Second-generation behavior results from an interplay of application programs, conditioning modules, and the robot's experiences. The behavior will change as experiences accumulate, and second-generation robots will learn to deal with circumstances not explicitly programmed. They can be trained or traumatized. They will seem no smarter than a small mammal outside the specific skills built into their application program of the moment, but they will *have* an outside, a general, nonspecific intelligence. They will have likes and dislikes, and learn to find the first and avoid the second. They will have a behavioral character. I'm certain many people will empathize and interact with second-generation machines as they do with pets, and the robots will respond.

The third generation adds a physical and psychological world-model to the older elements. The resulting behavior is likely to create a vivid impression of a conscious mind. A third-generation robot observes its world and builds a working simulation of it. The moment-to-moment state of the simulation is recorded so that the robot can later replay it, with variations, to learn. The simulator also runs in fast-forward so the robot can anticipate events and weigh alternative actions. An alternative's desirability is defined by the conditioning modules, as in the second generation. But the simulation can be used in another way, which eases the robot's interaction with people. Because it is composed of labeled objects and events, the simulation can be examined to answer questions about the robot's past and future actions and observations. "Where did you put the dust pan?" "What room are you going to clean next?" "Do we have any milk left?" A precedent for this kind of interaction with a world model is found in Terry Winograd's 1970 program SHRDLU, which answered questions and executed commands about a tiny simulated tabletop world with a dozen movable blocks and pyramids of various sizes and colors.

The simulation has psychological as well physical elements. "Did Mary seem worried when she left the house?" "Are you planning anything to cheer her up?" The psychological models are programs that interpret human behavior, especially words, and predict actions. Like other elements in the simulator, they are adjusted by the robot's experiences. Now, there may be other robots in the environment. Robots behave in too complex a manner to be modeled as physical objects. Their actions resemble human behavior, and properly tuned psychological models may be applicable to them. In that case a question, "Why did the vacbot not clean the mess at the head of the stairs?", might be answered by an observing third-generation robot with "It was afraid to get too close because it might fall down again, as it did last week." A third-generation robot has hardly any more insight into its own mechanism than it does of other robots. It has neither the information nor the processing power to simulate itself in detail. It can, however, apply psychological observations to its own behavior, as it does to other robots. Asked "Why did you put the flowers on the table this afternoon?" it might answer "Because I thought it would cheer Mary up." "Why do you want to cheer her up?" "Because I like it when she is happy." The latter response is a reference to the positive conditioning the robot receives for the actions leading to scenarios where Mary is judged to be happy. These signals come from a conditioning module keyed to psychological models, but the robot can't simulate its big conditioning suite and the complex effect it has on the myriad branch paths and recognizer weights in its control program. Instead, for the purposes of answering questions, it simulates itself with a generic human model, tuned to match its observations of its own behavior.

A third-generation robot's simulator-based mind is very concrete in its thinking and will seem naive and simpleminded in conversation. Yet it can honestly answer questions about its own motivations and feelings, plans and regrets. Insofar as one is willing to grant that it has beliefs, it believes itself to be conscious, just like other people!

Fourth-generation robots, more powerfully than humans, will be able to abstract generalities and derive subtle consequences from third-generation scenarios. Conversely, they will be able to reinstantiate abstract conclusions into concrete simulations, and perhaps abstract them again in a different way. Such machines will not seem simpleminded, but they could easily be incomprehensible unless their abstractions and presentation were carefully attuned to human patterns of thought. A fraction of their adaptability, modeling, and problem-solving power will surely be devoted to exploring the bounds of human comprehension, so they can tune themselves to speak—and perhaps think—in a way we can understand. Their conditioning suite will include negative modules for the situation that a human can't follow what they are saying.

The Burden of Consciousness

A third-generation robot's real-time simulation represents a memory of the past and an anticipation of the future. It enables robots to examine their actions and their effects as, or before, they occur, and to alter the behavior on the basis of the insights. If the simulation contains a modest representation of important features of the robot's interior, like battery charge, temperature, balance, even program state, then the simulator confers a certain amount of *self-awareness*. The robot's actions will be shaped, in part, by anticipation of how they will affect its inner state. Even apart from an interior model, there is an implicit indication of a self, as the center of the most detailed region of the simulation and the origin of all coordinate systems. The robot naturally sees itself as the center of its world. Most mammals probably have a nonverbal perceptual-motor consciousness something like this.

Some theories of human consciousness focus on the role of language, suggesting that our most pronounced sense of awareness results from a narrative we conduct within ourselves. Our language-generating faculties describe ongoing events, their logical consequences, and their emotional impact, while our language-understanding mechanisms listen and respond to the story, creating secondary reactions that may themselves be woven into the narrative. Human reasoning seems to be mainly a disciplined exercise of language, so talking to ourselves may be the only way to carry on extensive trains of thought.

The later robot generations will have language ability, but there seems little point in programming them to reason by talking to themselves. Unlike humans, they will have a much more direct, efficient, and powerful mechanism for the job. Fourth-generation robots conduct the equivalent of a narrative when they use their reasoning programs. The reasoners can relate simulator events to long-term goals, thus giving meaning to activities that have no immediate effect in short-term simulations. Such logical extrapolations could, for external amusement, be directed to the language-generation system. Like a demented sports announcer, this would voice a running commentary on the significance of every little thing that happened around the robot. More usefully, the value judgments implicit in the inferences could be converted into conditioning signals. These would subtly alter the robot's behavior in ways that enhanced its long-term objectives.

In robots, as in humans, consciousness has consequences. We look over our own shoulders and become our own meddling back-seat drivers! There are many ways to orchestrate the four-layer action/conditioning/simulation/reasoning control system of the fourth-generation robots. Some configurations will make a robot more thoroughly conscious than the average human and thus more likely to interrupt its own actions. It will be insightful, but often indecisive. Others will make it less conscious, more a "Machine of Action," likely to get its jobs done and also to blunder into messes. No doubt different tasks will benefit from different arrangements and emphases. Some oft-repeated jobs will be best done in a Zen-like state of unconscious competence. Others, unfamiliar or prone to unexpected failures, would best be supervised at many levels, each ready to be interrupted from above.

Eventually, thoughtful scientist-robots may develop powerful models that elucidate the tangled interactions of our kind of consciousness, allowing them to reliably create human-sized minds with particular properties. But by then their own minds, greatly expanded and improved by trial and error, will be far more elaborately organized than ours, and in operation probably as bewildering to them as ours are to us.

Fear, Shame, and Joy

To succeed in the competitive game of life, robots, like animals, will skate near the edge of the possible, often acting on too little information. For an organism that takes chances, the world is

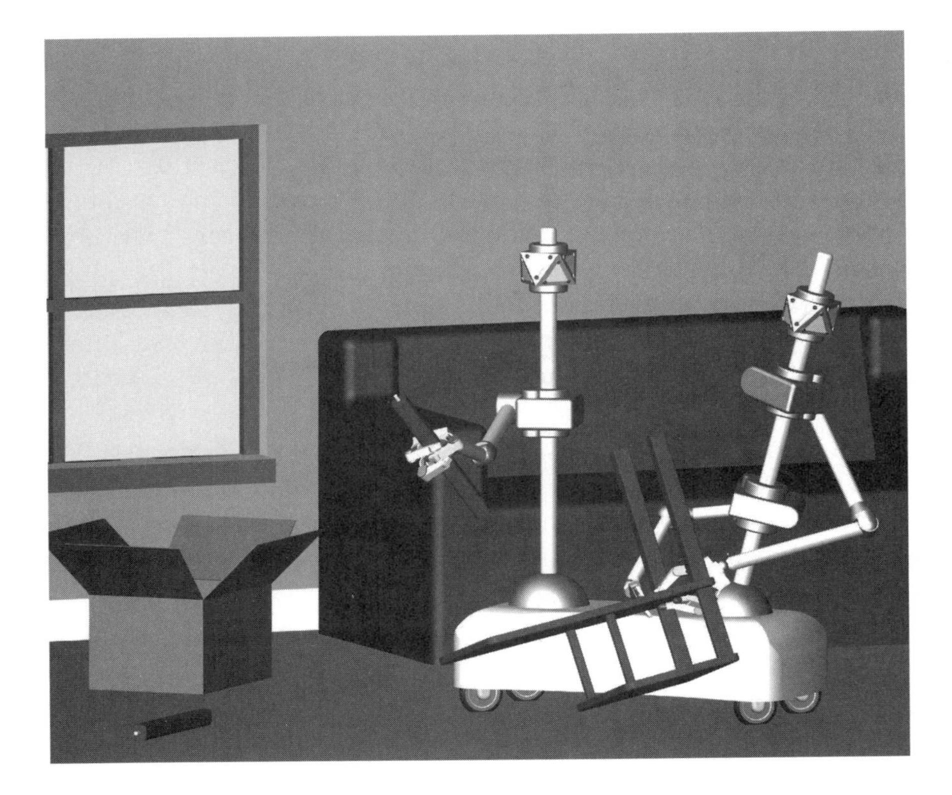

a tricky place, full of misleading clues and unexpected dangers. A first-generation robot navigates and identifies objects by statistically matching shapes and colors in its immediate sensory map against similar data stored in training sessions. Occasionally the robot may take a wrong turn, or misidentify something, and put itself in danger. There may be hints of trouble. If the robot unexpectedly enters a stairwell, the mapping software may indicate absence of a floor, or the navigation system may note unexpected surroundings. To forestall disasters, even first-generation robots should have ways of interrupting their application programs in case of danger.

Watchdog programs, like simple versions of the conditioning modules of the second-generation robots, might run simultaneously with application programs. They would look for signs of trouble, ready to activate appropriate response programs. Most emergency-response programs will probably make the robot be-

have more simply, singlemindedly, and energetically than the application programs they interrupt. If a watchdog program gets a hint of danger, it may switch the robot from seeking its destination at a measured pace to abruptly halting, increasing its sensing rate to verify and locate the problem, and slowly, carefully, backing away from the hazard, or, for some dangers, retreating as quickly as possible.

The robot would react in a way we recognize as fear when we see it in animals. Indeed, in simple animals like insects, the mechanism seems no more complex. Neural circuitry that recognizes potential dangers issues signals that activate motor ganglia that produce special defensive behaviors, interrupting other activities.

The fear reaction is more elaborate in higher animals and higher robots. In vertebrates, the first hints of danger produce a systemic adrenaline signal, accelerating metabolism, and redirecting blood from long-term operations like digestion to functions necessary for fight-or-flight, like perception, thought, and movement. Computer-controlled robots will probably make do with a digital signal broadcast to the relevant subprocesses. It will direct background processes not essential for the emergency to shut down and leave more computer capacity for the sensing and action routines.

Vertebrates are conditioned by frightening experiences to avoid whatever it was that led them to the situation. Similarly, in second-generation robots, the watchdog programs will also be negative conditioning routines, generating signals disinclining the robot in the future to do the things that led to the problem.

When they narrowly escape an accident, larger mammals, humans certainly, may vividly imagine what might have happened and how they might have acted differently. Perhaps they will also replay the scene in nightmares. Third-generation robots will certainly simulate alternative versions of any event that triggers a watchdog program, searching for variations that avoid the problem. They may do a few replays immediately after the emergency, and more later when they have time. Sometimes finding a safe alternative to a dangerous action may have such high priority that replaying the accident distracts the robot from other activities.

After a serious fright, after the panic subsides, human beings may review their life choices, ashamed of the way they have acted, and rationally decide to mend their ways, to avoid the behaviors that caused the problem. Similarly, a fourth-generation robot may employ its reasoning program to evaluate the effect of its short-term choices on its long-term goals and perhaps decide on a different approach to its problems, one that avoids the dangerous situations. On the other hand, if a human's choices have turned out particularly well in the long run, like resulting in a thriving, happy family, the human may experience joy, reinforcing the behaviors that produced the result. Similarly, a particularly fortunate fourth-generation robot may reason about the successful advances in its long-term projects and generate system-wide signals reinforcing its current behavior.

In general, robots will exhibit some of the emotions found in animals and humans because those emotions are an effective way to deal with the contingencies of life in the wide, wild world.

Love

Not every emotion found in humans makes sense in robots. Early generations of robots will be built in factories, where their equivalent of genomes will be found in design computers. As with worker ants, there will be no way for individual robots to reproduce, sexually or otherwise, so there will be no reason to install any programming for sexual behavior or feeling.

Yet there may be sound business reasons for giving robots a platonic love of their owners, similar to the loyalty bred into domestic dogs. In a third- or fourth-generation robot, the simulator may have a capacity to model the mental states of humans, and probably other robots, via "psychology" models. These would be trained like object recognition and interaction models, and kept tuned by subsequent experience. They would allow a robot to weigh the effect of its actions on the feelings of the humans it encounters. Just what effect it seeks depends on the conditioning modules it carries, but it should be possible to install a suite that strongly reinforces behavior that makes its owner happy. The robot would then use its inventive capacity (modest in the third generation, unbounded in the fourth) to devise activities to please its master. Perhaps it would carefully choose its work times so as not to disturb, perhaps it would prepare meals that especially pleased, perhaps it would set out favorite flowers, or perhaps it would find some way to bring extra income into the household. All these activities would be reinforced by the robot's conditioning system, which would provide the only reward it needed. Reciprocal consideration, or even thanks, would be unnecessary, except perhaps as a signal to its psychology modules that the owner was genuinely pleased. The robot would be exhibiting pure selfless, nonjudgmental devotion, the kind of love a good dog or a superb friend offers—what the Greeks and the Christians call "agape."

Of course, there is more to the story. In the social insects, workers seem to behave purely altruistically, exerting themselves to death for the good of the colony. Unlike solitary animals, who will disappear in future generations if they are not selfish enough to ensure their own survival and reproduction, worker insects are manufactured by the colony through the agency of the queen. Their genes will continue to exist only if the colony is successful, and their apparent altruism is actually selfishness on a larger scale. So it will be with selfless robots, who are made in factories. Robot models that please the customer will be recommended for future purchases, and manufacturers that make customer-pleasing robots will do well, and so "nice" robots will be made in increasing numbers. Surly or treacherous robots, on the other hand, will probably sell poorly and so will disappear, perhaps taking their manufacturers with them to extinction.

Customer selection may thus ensure that most robots that regularly interact with people will be very nice—nicer, in fact, than most humans, who have good genetic reasons to be selfish much of the time. But manufacturers of nice robots will be competing against each other. The very different shaping forces in that larger arena are the theme of the next chapter.

Anger

In animals, anger plays at least two roles. Certain situations are best solved by aggressive action. In others, competitive clashes can be won by bluster, by making it seem to an opponent that a battle is not worthwhile. The physiological symptoms of anger are similar to those of fear, but the probabilities are shifted in favor of the fight side of the fight/flight dichotomy.

It would probably be dangerous to give early generations of universal robots a capacity for anger, which might be triggered by a misunderstanding in their limited view of the world. Later robot generations may occasionally find some kinds of anger useful in interactions with irresponsible humans or irresponsible robots. For instance, suppose a fourth-generation robot is programmed to function as a security guard. On encountering an intruder, it might first request cooperation. If cooperation is not forthcoming, it may then issue threats. If threats fail, it may escalate to aggressive action. During the process, the robot would attempt to model and manipulate the mental state of the perpetrator, trying to induce cooperation. The first stages of interaction would be bluster, but the robot would be simulating future possibilities and considering those in which it, or the perpetrator, were damaged.

Normally, the possibility of damage would produce negative conditioning that would restrain the robot's actions. However, the security robot would contain a conditioning module that reacted to security threats by generating positive reinforcements for action. This would make the robot aggressive, causing it, perhaps, to ignore the danger and apprehend the miscreant. Doing so would incidentally reinforce its reputation for fierceness, making future encounters less likely to require violence. A fourth-generation robot given enough data about human behavior might be able to devise this plan on its own, given the job of guarding a place. It might consequently install in itself such an action-inducing conditioning module and consequently the ability to become angry.

Not ruled by unalterable inherited drives, fourth-generation robots will be more deliberate and adaptable in their behavior than humans. The ability to alter one's character has serious dangers, though. A robot could change itself in detrimental and irreversible ways—for instance by accidentally destroying its ability to think or making itself stubbornly resistant to change. Probably such changes should never be undertaken except in a "buddy" environment. Some friendly bystander, perhaps an unaltered copy of the robot itself, should be watching closely, ready to undo the change should it go awry.

Pain and Pleasure

Medieval scholastics imagined the physical world to be just a shadow of the spiritual world where God and human souls dwelt. Studying material things was considered a waste of precious time better spent in devotion—that is, communing with the actual spiritual causes of everything. Three centuries ago René Descartes upset this view with a radical alternative. Having observed the likes of hydraulically activated puppets, clockwork ducks, and the imaging of bovine eyes, he suggested that the body was just a complex machine, probably run by hydraulic pressure of the bloodstream. Lacking a mechanical model for thought, he retained part of the medieval idea. The mind was a spiritual entity that interacted with the mechanical body through the agency of the pineal gland, the centrally placed "third eye" in the brain. The pineal's view of the physical world is blocked, so it must be for seeing the ethereal realms instead. If he were working today, Descartes might well have found, in computers, a material model for mind and become a thoroughgoing materialist. But, alas, there were no computers in the seventeenth century.

Cartesian body-mind dualism dominated Western philosophy and science until the early twentieth century, even among those who had rejected its religious roots. This century, scientific thinking was overwhelmed by Darwin's nonmystical explanations for the development of complex life and by detailed studies of the brain's anatomy and function. There were ever more reasons to suspect that even the most mysterious mental phenomena had physical causes. Dualism, the "ghost in the machine," became unfashionable in philosophical circles with the arguments of Gilbert Ryle in the 1930s and 1940s. But the idea did not die.

It seems obvious to many that a dead mechanical process cannot by itself give rise to our own mental experience. The idea of having a special soul also appeals to a tribal instinct for identity, uniqueness, and superiority. It was the starting point of Descartes' analysis, "I think therefore I am." It still appears in cryptic form in the arguments of critics of materialistic science in general and thinking machines in particular. Hubert Dreyfus, a philosopher at Berkeley, argues that computers may imitate the conscious surface of thought, but can never capture the ineffable intuitive subconscious. His colleague John Searle says that computers may *simulate* thought, but will never actually think meaningfully, just as computers may simulate a liver, but can never secrete actual bile fluid, thus implying that thought and meaning are unseen substances. Roger Penrose argued a few years ago that the brain achieves consciousness and gains access to a platonic realm of infinite mathematical truths through gravitational collapse of the quantum wave function in individual neurons.

How can a dead mechanical process produce our vivid mental life? Neuroscientists and philosophers of mind have begun to construct answers from the accumulating, but still insufficient, experimental evidence. Daniel Dennett's book, *Consciousness Explained,* explains our interior mental lives as a kind of illustrated story, told in the activation of sensory and motor circuitry as well as language. The sensation of the color red, for instance, is a symphonic note involving reflexes for blood, berries, and stoplights, and possibly the word "red" and its referents. It is very different from the activations for "green." The components of the sensation happen independently, and are coordinated only after the fact, as we recount the immediate past to others or ourselves. In this retelling we are generally not very good witnesses. The records seem altered for the sake of a better story, as were histories in Stalinist Russia. A minor example involves our perception of simultaneity of a visual and a tactile stimulus. Complex visual processing takes many milliseconds longer than tactile sensing, so a simultaneous flash and touch register at the cerebral cortex at different times. Yet we perceive the two as simultaneous. In brain experiments, if the relevant areas of the cortex are simultaneously stimulated electrically, the induced visual sensation appears to precede the tactile one. The memory of the two arrival times seems to have been edited to compensate for differing transmission delays. Extreme conditions such as hysterical blindness or hallucinations are more dramatic warnings that our innermost experiences may not be quite everything they seem. Our subjective experience is an edited story. If our memories are untrustworthy, how can we be sure we really experienced what we thought we did?

Those who support the traditional view—that subjective experience is the ultimate reality—point to vivid experiences like intense pain as evidence. How can something that makes us struggle so, which leaves us so traumatized afterwards, be less than utterly real?

Consider what might pass for pain in a fourth-generation robot. The sensation would be associated with dire emergencies that need immediate action. Suppose sensors in the structure of the robot register that the metal is starting to *bend*! Severe damage is imminent! A well-designed robot has a watchdog program for this case that immediately activates a routine that tries to pull the robot away from the excessive force. It would be ready to use dangerous emergency power levels if necessary. It might even call for help. Maximum negative conditioning signals are sent out, greatly inhibiting ongoing behaviors. If the problem does not go away, the negative conditioning would cause the robot to switch behaviors repeatedly by suppressing existing actions, making their alternatives more likely. After some amount of thrashing around, the robot might extricate itself. Asked what happened, the robot will replay a simulation of the event and note the intense negative conditioning. Its reasoning system may observe the replay and concoct the explanation: "I was shifting some boxes when a heavy one fell on me. My frame was being bent, it was a *very bad* experience the whole time. Even thinking about it is unpleasant. I tried so hard to get out from under the box that my drive motors overheated, but that was minor in comparison. I'll never handle boxes that way again." "Very bad" is a generic reference to a state of extreme negative conditioning. Stories and simulated replays are all that remain after the event. If the simulator is configured to limit the intensity of negative conditioning on replays, to minimize the chance of inducing a debilitating phobia, the memory will not be as vivid as the original event. Perhaps the reasoner will have stored a dispassionate account of the original intensity and so will be able to comment abstractly on the difference between experienced and remembered pain.

Pleasure, of course, would be an analogous experience associated with intense positive conditioning. Perhaps the robot will be configured to experience it if it accomplishes its chores extraordinarily effectively, or makes its owner exceptionally pleased.

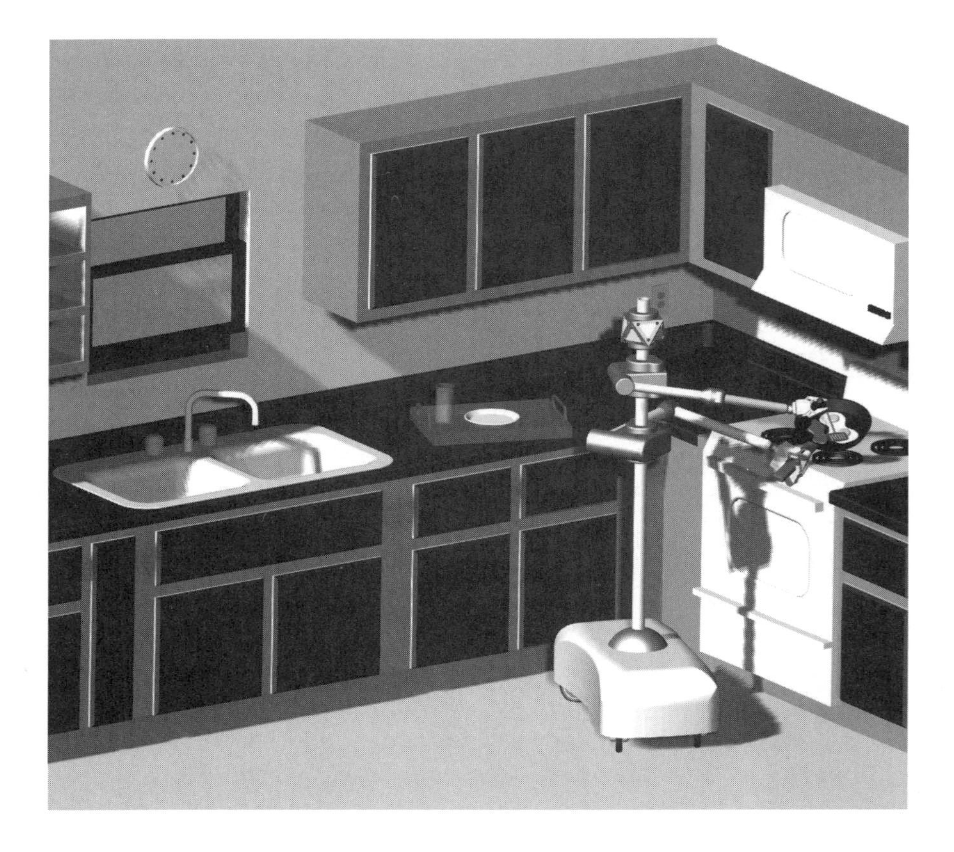

Superrationality

The speculative evolution of robot minds above emphasized parallels with natural minds. But computers and brains have different relative strengths. Fourth-generation robots will be much better reasoners than human beings. They will make inferences at least a million times as fast and have a million times the short-term memory. Reasoning is computationally universal. It can simulate any other computation, and so could, in principle, do the job of the world modeler, the conditioning system, or the application program itself.

Simulating a control system in a reasoner would be slower than running the controller directly on the computer. But the reasoner is also able to abstractly examine the simulation and devise shortcuts for the complicated operations. By optimizing itself constantly, a reasoner-based controller might ultimately become faster, possibly much faster, than a direct program. It might be able to peer much farther into the future and consider a wider range of contingencies.

It might be possible to construct totally reasoning-based robots. In them, even the smallest activities, instead of being inflexible conditioned reflexes, would be carefully planned steps toward long-range purposes. The robot's every twitch would be like a move in a chess game. A super-rational robot might not act very differently most of the time from a conventionally organized machine. But once in a while some of its actions would conspire to produce a spectacular coincidence that accomplished some end, like a pool hustler's twenty-cushion trick shot that produces a bewildering flurry of activity and sinks ten balls. On the other hand, long contingent chains are extended opportunities for something to go awry in the noisy physical world. Even super-rational robots might have to content themselves with nudging events along bit by bit, correcting for unpredictable disturbances as they happen.

Mind Children

The fourth robot generation, and its successors, will have human perceptual and motor abilities and superior reasoning powers. They could replace us in every essential task and, in principle, operate our society increasingly well without us. They would run the companies and do the research as well as performing the productive work. Machines can be designed to work well in outer space. Production could move to the greater resources of the solar system, leaving behind a nature preserve subsidized from space. Meek humans would inherit the earth, while rapidly evolving machines expanded into the rest of the universe.

The development can be viewed as a very natural one. Human beings have two channels of heredity. One is the old biological variety, encoded in DNA. The other is cultural information, passed from mind to mind by example, language, books. and, recently, machines. At present, the two are inextricably linked. The cultural part, however, is evolving very rapidly and gradually assuming functions once the exclusive province of the biology. For most of our species' history, there was less data in our cultural heritage than in our genomes. But in recent centuries culture has overtaken genetics and now our libraries hold thousands of times as much information as our genes. Barely noticing the transition, we have become overwhelmingly cultural beings. Ever less biology animates ever more culture.

Given fully intelligent robots, culture becomes completely independent of biology. Intelligent machines, which will grow from us, learn our skills, and initially share our goals and values, will be the children of our minds. The next chapter suggests how we parents can gracefully retire as our mind children grow beyond our imagining.

5

The Age of Robots

A thousand centuries ago, the world was fully automated. Our ancestors were supported by the maintenance-free, self-operating machinery called Nature. But, in an Adamic bargain predating Faust, they meddled with the mechanism. By tilling and planting, they magnified the machinery's productivity but trapped themselves in a routine of heavy, unpleasant labor. Anthropologists were surprised by ergonomic analyses showing that citizens of advanced societies worked far longer and harder for their living than primitive hunter-gatherers. Millennia ago, many American Indian tribes were aware of the disparity and consciously chose to be sparse nomadic hunters rather than sweating farmers. Meanwhile, the agricultural civilizations burnt their bridges. Populations grew, and became dependent on ever more elaborate social and physical inventions. Prominent among them were ways of moving civilization's unnatural burdens to draft animals, slaves, and machines. For most of history, these developments, while increasing society's overall vigor and control of nature, merely shifted the toil to other backs or slightly different chores. Only in the last few centuries has labor-saving machinery begun to overtake labor-creating expansions of activity and so begun to diminish the total amount of work required of civilized humans. Despite occasional brief setbacks, the trend is accelerating. After millennia lived by the sweat of our brow, we are at last providing more and more with less and less work.

In all recorded history, until this century, nearly every civilized person was a farmer. We were consumed with raising our food. In spare moments we maintained our shelter and clothing. Institutions like slavery, feudalism, and capitalism lifted a small minority out of the drudge on the backs of the rest. A few of the few used the gift to seek substitutes for hard work. It was a slow process that ultimately resulted in the Industrial Revolution. In this century industrialization—biological, chemical, and organizational innovations, but mainly giant machines that do the work of many humans but are controlled by fewer—has finally ended most agricultural labor. In the United States a few percent of the population produces surplus food for all. Displaced farm workers moved into manufacture where, besides filling many new needs, they made the farm machines. But just a few decades later, more advanced machines do most of the manufacturing. Displaced factory workers have moved into offices where, besides filling many new needs, they devise and direct the factory machines. But while that transition still reverberates, even more advanced machines are edging into engineering, management, and customer relations work. Humans have been upwardly mobile in the jobs pyramid, but will soon be squeezed out of the apex!

It is sometimes said that mass manufacturing ended the role of the craftsman, who was replaced by heavy machines stamping out identical parts. In fact, the craftsmanship moved into the engineering department, where products are designed and their means of manufacture devised. Engineering is a thought-, labor-, and creativity-intensive endeavor. In the late 1980s it began to be massively automated.

As tractors and combines amplify farmers, computer design workstations amplify engineers. A designer clicks a change in a part and immediately sees its effect on the rest of the design. In the 1970s such an investigation conducted on drafting boards might have engaged junior engineers for weeks. The effects on productivity and employment may never be precisely measured, but enterprises as diverse as plastic injection-mold designers and architectural firms reported tenfold throughput increases by the same staff when design and accounting computers replaced manual methods. New integrated circuits and computers, though a hundred times as complex as last decade's, appear nine months after conception rather than in three years. Increasingly powerful computers are enabling new software to do even more of the designer's job. Some tools automatically sort through design alternatives, looking for configurations that best meet cost and performance specifications. Others simulate circuitry, software, and mechanical structures to evaluate products before they are built. Linked directly to manufacturing machines in the factory, design systems are on the verge of allowing a handful of engineers to make a cornucopia of excellent products, much as a few farmers in the air-conditioned cabs of powerful combines can harvest enough wheat to feed a city.

Meanwhile, in the office, layers of management and clerical help are evaporating. From the start, computers displaced workers in number-crunching jobs like accounting. Word-oriented office tasks were untouched until recently. In the 1970s the Stanford Artificial Intelligence Laboratory, with over fifty faculty, staff, and students, prided itself in needing only one secretary. Using the lab's 1-MIPS time-shared computer, which seemed powerful at the time, everyone was expected to edit, format, and print their own documents, to issue reminders and memos via computer mail, and to do accounting and scheduling with specially written programs. The lab also had among the first engineering design programs. It attracted so many excellent programmers (who called themselves "hackers") that new capabilities appeared almost daily. Today, offices everywhere are more automated than SAIL was. New generations of software, doing the data-gathering and decision-making functions of whole departments, will cut staff even more deeply.

Once upon a time, raconteurs, dancers, and singers in every village were a tribe's historical memory, repositories of its sagas. Then a radical new technology—writing—usurped their main function. One storytelling could now be multiplied by rote transcription to reach thousands across space and time. Most performers became mere entertainers. Printing, the phonograph, film, radio, and television, by increasing the amplification factor to millions, made even entertainment a fantastically competitive business, needing only a few originators. Computer technologies are eroding what jobs remain. Most commercial music is generated by computerized music synthesizers, not musicians. Ultrarealistic 3D computer animations are supplanting real objects and lifeforms in television and film. Soon they will be able to replace principal actors.

The employment lowlands of number crunching and paperwork have long been submerged under superior computer labor. The flood has now reached the base of the mountain ranges of human physical and social skills. Telephone companies long ago replaced most of their operators with automatic switches. Now voiced queries are being handled by speech-recognizing computers. In the 1980s machines began to sort most mail for the U.S. Post Office, but obscured or handwritten addresses were shunted to humans. Now trainable text-reading machines are taking over. In the same decade expert systems, made of painstakingly handcrafted decision rules, began to advise many businesses, configuring computers, prospecting for oil, authorizing credit-card purchases, even making medical diagnoses and suggesting treatment. In the 1990s they are being replaced by second-generation systems that learn many of their skills automatically. Robots cannot yet handle material objects with human flexibility, and are useful only in highly structured environments like factories. In the broader world, intricate manual work is still in hands of flesh.

Mechanization and automation have concentrated human work. Devices like earthmovers, combines, television stations, automatic milling machines and office computers permit a few skilled operators to replace a greater number of humbler laborers. The machines work prodigiously, but require human direction and assistance. Some see this as the natural, desirable, and inevitable relation between humans and machines, implicitly drawing a line between human and machine skills. But the boundary is not static. Machines cannot yet match human motor control, judgment, or emotional empathy. But they are advancing rapidly on all fronts, and the human component of competitive business is shrinking. A telephone, a cash register, or a milling machine requires a skilled operator for each transaction. Voice mail, a bank machine, and a computer-integrated factory may run autonomously for days. Rising productivity is a business imperative as long as customers choose better goods at lower prices. Output per worker must increase, and so the amount of essential labor decreases.

This century's labor movement spread productivity's wealth by giving most workers higher pay for shorter hours. World War II temporarily halted the trend by diverting laborers and production to service the conflict, then reconstruction. But, by the late 1950s, economic conferences and publications began to worry about the problem of excess leisure time. In the United States aggressively nurtured consumerism and a fabulously expensive Cold War absorbed surplus productivity. In parts of Europe, on the other hand, the work week shrank to thirty-five hours, vacations grew to several months, and chronic unemployment reached 20%.

Advancing automation and a coming army of robots will displace labor as never before. In the short run this threatens unemployment and panicked scrambles for new ways to earn a living. In the medium run, it is a wonderful opportunity to recapture the comfortable pace of a tribal village while retaining the benefits of technological evolution. In the long run, it marks the end of the dominance of biological humans and the beginning of the age of robots.

The Short Run (Early 2000s)

The Industrial Revolution gathered steam two centuries ago. It destroyed cottage industries and concentrated wealth in the hands of factory owners—the capitalists. Millions of displaced home workers competed for too few jobs tending the new machines. It took difficult political readjustments to spread the benefits of cheaper, more plentiful goods. Gradually laborers' hours were reduced from the traditional fourteen per day. More workers were then needed and salaries were bid up.

Although it increases communal wealth, each increment in automation threatens a similar unpleasant transition, as it displaces one group of workers with fewer doing different tasks. If the new tasks require common skills, mass competition for the few jobs drives down salaries. If they need rare skills, scarcity encourages high pay and long hours. Either way, some work excessively while others are jobless. It takes slow changes in the social contract and education to level the load.

Even though work hours will decline, they cannot be the final answer to rising productivity. In the next century inexpensive but capable robots will displace human labor so broadly that the average workday would have to plummet to practically zero to keep everyone usefully employed. Already, much labor services questionable ends—gargantuan government bureaucracies, cosmetic medicine, mass entertainment, and speculative writing, to give a few examples. In time almost all humans may work to amuse other humans, while robots run competitive primary industries, like food production and manufacturing.

There is a problem with this picture. The "service economy" functions today because many humans willing to buy services work in the primary industries. They return money to the service providers who in turn use it to buy life's essentials. As the pool of humans in the primary industries evaporates, the flow of primary money to the service sector will decline. Efficient, no-nonsense robots have no need for frivolous services. Money will accumulate in the industries, enriching just the few people still associated with them. Cash will become scarce among the service providers. Primary product prices will plummet, reflecting both the reduced costs of production and the reduced means of the consumers. In the ridiculous extreme, the service providers would run out of money and the robots would fill warehouses with essential goods that human consumers could not buy.

Not all individuals involved in productive enterprises actually work there. Stockholders, having once contributed capital, may collect dividends indefinitely from a thriving enterprise. Workers can be replaced by automation, but owners remain until they sell out, and may, for a while, be the major conduit for spending money. An analogous situation existed in classical and feudal times where an impoverished, overworked majority of slaves or serfs played the role of robots and landowners played the role of capitalists. Between the serfs and the lords a working population struggled to make a living from secondary sources, often by performing services for the privileged. A prestigious and prosperous minority of commoners sold high-quality goods and services directly to the gentry (as in the proud line still seen in Britain, *By Appointment to Her Majesty*). The majority lived more modestly, from trade with other townspeople.

It is unlikely that a future majority of service-providing "commoners" with more free time, communications, and democracy than today would tolerate being lorded over by a dynasty of non-working hereditary capitalists. They would vote to change the system. The trend in the social democracies has been to equalize income by raising the standards of the poorest as high as the economy can bear. In the age of robots, that minimum will be very high. In the early 1980s James Albus, head of the automation division of the then-National Bureau of Standards, suggested that the negative effects of total automation could be avoided by giving all citizens stock in trusts that owned automated industries, making everyone a capitalist. Those who chose to squander their birthright could work for others, but most would simply live off their stock income. Even today, the public indirectly owns a majority of the capital in the country through compounding private pension funds.

Capitalism's End

Sooner rather than later, ownership may become as unreliable a source of human income as was robot-displaced labor. In a fluid and fiercely competitive economy, companies that squander resources by paying owners will be outdone and driven out of business by those that reinvest everything in productive operations and development. Like humans pushed out of labor markets by cheaper and better robotic workers, owners will be pushed out of capital markets by much cheaper and better robotic decision makers.

The evaporation of ownership will end capitalism, but capital enterprises will thrive as never before. Some companies will die, but others will grow. Those that grow especially well will be induced to divide by antitrust laws. Some companies may decide to cooperate in joint ventures that produce new enterprises that are a mix of the parent firms' goals and skills. With no return on investment in a hypercompetitive marketplace, the effort may kill the parents. But, if the offspring grows and divides, the parents' way of thinking may become more widespread than ever. The dynamics of capitalism will be replaced by the dynamics of biological reproduction. The ultimate payoff for success in the marketplace will no longer be monetary return on investment, but reproductive success.

Biological species almost never survive encounters with superior competitors. Ten million years ago, South and North America were separated by a sunken Panama isthmus. South America, like Australia today, was populated by marsupial mammals, including pouched equivalents of rats, deers, and tigers. When the isthmus connecting North and South America rose, it took only a few thousand years for the northern placental species, with slightly more effective metabolisms and reproductive and nervous systems, to displace and eliminate almost all the southern marsupials.

In a completely free marketplace, superior robots would surely affect humans as North American placentals affected South American marsupials (and as humans have affected countless species). Robotic industries would compete vigorously among themselves for matter, energy, and space, incidentally driving their price beyond human reach. Unable to afford the necessities of life, biological humans would be squeezed out of existence.

There is probably some breathing room, because we do not live in a completely free marketplace. Government coerces nonmarket behavior, especially by collecting taxes. Judiciously applied, governmental coercion could support human populations in high style on the fruits of robot labor, perhaps for a long while. In the United States, the Social Security system offers an evolutionary route to this end. Social Security was originally presented as a pension fund that accumulated wages for retirement, but in practice it simply transfers income from workers to retirees. The system will probably be subsidized from general taxes in the coming decades when too few workers are available to support the retiring post-World-War-II baby boom. Incremental expansion of such a subsidy would let money from robot industries, collected as corporate taxes, be returned to the general population as pension payments. By gradually lowering the retirement age, most of the population would eventually be supported. The money could be distributed under other names, but calling it a pension is meaningful symbolism. Social Security pension payments begun at birth would subsidize a long, comfortable retirement for the entire original-model human race.

The Medium Run (Around 2050)

What happens to people when work becomes passé? Existing retirement communities are probably too sleepy to be a good model. Most of the individuals there have completed their life's work and are of declining vigor and health. Better examples may be the richest Arabian petro-kingdoms where oil-bought foreign labor plays the role of total automation. In a tradition of tribal sharing shaped by a sparsely-furnished nomadic past, Kuwait, Saudi Arabia, and the United Arab Emirates have managed to spread great wealth broadly among the citizenry in a single generation. Free health care and education, and undemanding government jobs, or outright welfare, secure life's needs. Life expectancies and literacy rates are among the world's highest. Comfort and security mute the stresses of civilization, including the tension between circumscribed Islamic values and the liberties of a wealthy world culture. The societies produce both world-class achievers and criminals, but on average show less driven urgency than many industrialized nations. Most of their citizens seem happy to simply live their lives, and stability is endangered only by neighboring countries, where impoverished majorities are less content with the status quo. On a smaller scale, wealthy families worldwide often produce generations of content, even smug, heirs, with a few exceptions to titillate the tabloids.

Contrary to the fears of some enmeshed in civilization's work ethic, our tribal past prepared us well for lives as idle rich. In a good climate and location the hunter-gatherer's lot can be pleasant indeed. An afternoon's outing picking berries or catching fish—what we civilized types would recognize as a recreational weekend—provides life's needs for several days. The rest of the time can be spent with children, socializing, or simply resting. This is the life Davi Kopenawa, the Yanomami spokesman introduced in the first chapter, was begging to protect from civilization's mania.

Of course, our ancestors also had to survive hard times, and evolution bequeathed us the capacity for desperate measures, including hard work. Civilization turned that extremity into everyday normality, and now stress is the leading cause of disease and probably triggers some of the ugliest aspects of tribalism. In primates, overpopulation is a common reason for group distress, as nature-provided food and shelter falls short. To survive, a strong tribe may chase away or exterminate a weaker neighbor, or drive out or otherwise eliminate some of its own members—maybe those who smell, look, sound, or act differently. Sometimes stressed individuals become accident- or disease-prone and die spontaneously, improving the prospects for their relatives. Similar considerations may regulate the prevalence of nonreproductive behaviors like homosexuality.

City life, absurdly crowded and stressful by tribal-village standards, may inappropriately trigger unconscious overpopulation reflexes. The self-destructive emotional vehemence of ethnic strife worldwide hardly represents rational self-interest. Modern tensions may subside when robot labor gives us a work-free life and the freedom to abandon the cities. It will be harder to muster battle fervor against minorities in populations that exist in a state of luxurious lassitude.

Ultraconservative Switzerland may be a hint of things to come. Government and commercial institutions perfected in centuries of peace (interrupted only briefly by Napoleon) have given Switzerland unmatched prosperity, stability, and security. Most Swiss citizens work, but they do so comfortably. Generous government welfare, and Italian immigrant labor, has lessened the desperation that forces workers elsewhere into unpleasant jobs. Multi-ethnic, multi-religious, multi-language Switzerland is made of twenty-three fiercely independent cantons each with its own traditions and history. Yet comfortable prosperity has allowed it to peacefully endure the most severe internal quarrels—for instance, the political fury between German and French factions during the First World War. The average Swiss citizen may resist most major changes (why ruin a good thing?), but stodgy Switzerland produces world-class contributors in all fields. Although it gives everyone the opportunity to excel, it lacks the social trauma that drives some other countries. Few Swiss would prefer it otherwise.

Unemployment

Many trends in industrialized countries lead to a future where humans are supported by machines, as our ancestors were by wildlife. Technology and global competition are gradually depopulating businesses. Flexible automation is displacing labor in food production and manufacture. Communicating computers are replacing clerks, secretaries, and managers in offices. Jobs that still require human labor are drifting to the homes of computer-equipped telecommuters. Reduced office staffs need less catering, janitorial, and maintenance support.

Gradually, machines capable of policy-making, public relations, law, engineering, and research will replace telecommuters. Advanced robots will displace technicians, janitors, vehicle drivers, and construction crews. Underemployed populations will be increasingly likely to vote themselves income from taxes on labor-free but superbly productive industries. Nor are less developed countries likely to be left behind. They can attract industries simply by offering competitive tax rates. Production today depends on an educated workforce, but fully automated companies require only space and raw materials.

Western democracies may turn into lazier Switzerlands. Big cities will lose their economic advantages and may begin to evaporate. Individuals, linked by worldwide communications and served by personal robots, can scatter to areas offering more elbow room. Nations may become less important, as taxes on local robot industries supply all human needs. The civilized world may return to a comfortable tribalism after a five-millennium detour into organized civilization. Countries with traditional tribal structures may simply stay that way, building on their ancestral customs, leapfrogging urbanization altogether. Developed countries may foster untraditional tribes with customs and beliefs more bizarre than anything today.

Wealth and technology will let tribalism express itself in novel ways. Over the last two decades, computer networks have hosted small communities whose members happen to be distributed around the world. "Usenet," started in the 1970s as an unofficial way for computers to exchange electronic mail by telephone, had grown by the 1990s to about ten million subscribers. It carried about three thousand specialized discussions on every conceivable topic, some burdened by fifty full-page messages every day. Regular contributors to particular "newsgroups" soon begin to recognize one another and develop characteristic interactions, likes, and dislikes. They formed factions that praised, recruited, condemned, and ostracized. When a newsgroup grew too large and noisy, specialized subgroups were formed, reducing the original group's population.

Usenet still exists on the Internet, but even though it has grown to thirty thousand newsgroups, it has been mightily eclipsed by the rapidly evolving World Wide Web, which is well on the way to absorbing everything and everybody.

The capacity of the world network will continue to grow, and new abilities, such as unobtrusive language translation, intelligent search, and helpful artificial personalities, will become integral. "Tribes of common interest" will share not just text, audio, and video, but full sensory environments. Tribal lands will exist in the minds of computers, in number, variety, accessibility, and properties impossible in the physical world. That is a topic for the next chapter.

While computer simulations create entirely new worlds, robots will transform physical life. Today manufactured items are difficult to make and thus relatively rare and expensive. We expend great effort in acquiring and defending goods. Our homes are fortified warehouses of our possessions. Stockpiling will be less appropriate amid robotic abundance. Why hoard fruit in an orchard? Conventional manufacturing methods—molding, casting, milling, assembly—can be robotically orchestrated to make new items fairly quickly. Even better, robotic accuracy and patience can build up solid objects by precisely "painting" various materials, layer upon cross-sectional layer. Such new approaches, refined to molecular resolution, will produce arbitrary solid objects from computer descriptions. Humans may be able to live in uncluttered spaces, in ecological preserves, if they choose. Robots could construct needed items, even food or housing, on the spot, or assemble them from nearby caches. Items no longer needed could be disassembled back into raw materials. Human surroundings would be tailored to the changing whims of their inhabitants, varying from the finest palaces to the rawest ecologies.

The most visible consumer items may be the "customer service" type of goods- and services-providing robot. They will come in human-pleasing sizes, shapes, accessories, and programming and may lurk unobtrusively until needed. Behind the scenes, a vast multitude of inconspicuous robots will work tirelessly to actually build, maintain, and operate everything.

That Is the Law

Service robots and heavier items will be manufactured by utilitarian robots that process energy and raw materials in bulk and conduct heavy engineering, exploration, and research. Molded by the constraints of the physical world rather than human tastes, these worker machines are likely to become ever more varied in size, shape, and function. They will form a growing ecology of artificial life that will eventually surpass the existing biosphere in diversity. Robot companies will be born from existing human firms, in familiar industrial settings near population centers, but competition will soon drive them to cheaper sites, perhaps locations that people find too hot, too cold, too dry, too poisonous, too far underground, or too remote. Robot companies' behavior will be shaped by future editions of existing laws, as well as by taxes and consumer whims.

Existing laws give incorporated entities some of the rights of persons, especially the rights to own property and make contracts. Lawbreaking corporations can be punished by fines, operating restrictions, or dissolution. Corporations do not have the right to life; they may legally be killed by competition or legal or financial actions. They do not have the right to vote on the laws that govern and tax them. It is a matter of life or death for the biological human race that the latter restrictions be strictly applied to evolving machine intelligences. Humans have a chance of retiring comfortably only if they themselves set corporate taxes, and all other corporate laws, in their own self-interest. The machines will be dangerously powerful physically and mentally, but can probably be constructed to be law-abiding. Some debate is inevitable, but there should be few qualms about keeping even very superior thinking machines in disenfranchised bondage. It takes force, indoctrination, and constant vigilance to counter inherited needs and motivations to enslave a human. Robots, on the other hand, do not have natural survival or any other instincts. Every nuance of their motivation is a design choice. They can be constructed to enjoy the role of servant to humankind. Nature itself provides examples of individuals motivated to serve more than survive, in the selfless worker castes of social insects and self-sacrificing mothers of all species.

The primary job of humanity in the next century will be protecting its retirement benefits by ensuring continued cooperation from the robot industries. The robots will present a moving target, but the instruments of control will also grow in power. Laws should demand that robot companies be built securely "nice" in the first place. Additional laws should require that niceness be enforced if necessary.

Corporate intelligences may be constructed like last chapter's fourth-generation robot minds. Immensely powerful reasoning and simulation modules will plan complex actions, but the desirability of possible outcomes will be defined by much simpler positive and negative conditioning modules. The conditioning suite will shape the character of the entire entity by defining its likes and dislikes. A company will be as unlikely to do something triggering strong negative signals as a human would be to thrust an arm in a fire. If the super-rational style of intelligent machines works out, robot character may instead reside in an elaborate body of axioms, contrived to be inconsistent with prohibited behaviors.

We voters should mandate installation of an elaborate analog of Isaac Asimov's "Laws of Robotics" in the corporate character of every powerful intelligent machine. It would take the form of an entire body of corporate law, with human rights and antitrust provisions along with appropriate relative weightings to resolve conflicts. Robot corporations so constituted would have no desire to cheat. They might sometimes find creative interpretations of the laws, which will consequently require constant tuning by vigilant humanity to safeguard their intended spirit.

Internalized laws, properly adjusted, could produce extraordinarily trustworthy entities, happy to die to ensure their legality. Even so, accident, unintended interactions, or human malice could occasionally produce a rogue robot or corporation, with superhuman intelligence and illegal goals. "Police" clauses in the core corporate laws that induce legal corporations to collectively suppress outlaws would mitigate the danger. Law-abiders would withhold services or, if necessary, use force to stifle lawbreakers. The laws should have antitrust provisions to prevent any corporation from growing too large to be suppressed in this way. The antitrust provisions would limit collusion between companies and cause overgrown corporations to divide into competing entities, ensuring diversity and multiplicity. Because dangerous robotic wildlife may eventually evolve in places beyond the reach of the law, the police clauses should also include provisions for a coordinated planetary defense against external threats.

Market Forces

Corporations live by building and maintaining physical assets that generate income to pay their expenses. In our proposed future, nothing prevents humans from increasing their Social Security income by raising corporate taxes. Taxation will surely be industry's biggest expense, and corporations will live or die by their ability to raise money to pay their taxes. Social-Security-rich humans will be the main repository of money, and robot corporations will have to compete mightily with one another to supply products and services that humans want to buy.

Like basic food in today's developed countries, common manufactured goods in the next century will be too cheap and plentiful to be very profitable. Most companies will be forced to continually invent unique products and services in a race against competitors to attract increasingly sophisticated (or jaded) human consumers. Automated research, as superhumanly systematic, industrious, and speedy as robot manufacturing, will generate a succession of new products, as well as improved robot researchers and models of the physical and social world. The likely results will exceed the dreams of science fiction. There will be robotic playmates, virtual realities, and personalized works of art that stir the emotions like nothing before, medical solutions for every physical, mental, or cosmetic whim, answers to satisfy any curiosity, luxury visits to almost anywhere, and things yet unimagined. The existence of an astronomically increasing variety of consumer choices will accelerate the divergence of human tribes. Some may choose a comfortable imitation of an earlier period, as the Amish today. Others will push the human envelope in wisdom, pleasure, beauty, ugliness, spirituality, banality, and every other direction. The choices made by diverse human communities will shape robot evolution. Only companies able to devise services of interest to the customers will generate enough income to survive.

Humans too will be shaped by the relationship. Robot services will be inexpensive, but not free, and income will be finite. Corporations will operate globally, but taxes will increasingly be assessed and redistributed on a tribal scale. Tribes that tax too heavily will drive away the corporations and so eliminate their revenue. Like tribes of the past that overburdened their ecology, they will learn to be modest in their demands on the land. More subtly, corporations struggling to appeal to consumers will develop and act on increasingly detailed and accurate models of human psychology. The superintelligences, just doing their job, will peer into the workings of human minds and manipulate them with subtle cues and nudges, like adults redirecting toddlers.

Being Human

Prosperity beyond imagination should eliminate most instinctive triggers of aggression, but it will not prevent an occasional individual or group from deciding to make mischief. Normal human actions will not be very dangerous in a world where cheap superhuman robots function as sleepless sentries, prescient detectives, fearless bodyguards, or, failing in the former, physicians able to resurrect dead people from fragments or digital recordings. But there is no limit to the troublemaking potential of humans with unconstrained access to robotic capabilities. For everyone's safety, the laws must prevent corporations from selling the means to make mayhem. Perhaps every powerful device sold to a human will include a fully intelligent interlock, or a robotic watchdog, that prevents it from being used for nefarious purposes.

Besides buying dangerous things, there is the possibility of *becoming* a dangerous thing. Humans can be enhanced by both biological and hard robotic technologies. Such present-day examples as hormonal and genetic tuning of body growth and function, pacemakers, artificial hearts, powered artificial limbs, hearing aids, and night-vision devices are faint hints of future possibilities. *Mind Children* speculated on ways to preserve a person while replacing every part of body and brain with superior artificial substitutes. A biological human, not bound by corporate law, could grow into something seriously dangerous once transformed into an unbounded superintelligent robot. There are many subtle routes to such a transformation, and there will be those who find the option of personally transcending their biological humanity attractive enough to pursue clandestinely were it outlawed. There could be very ugly confrontations when they are eventually discovered.

On the other hand, without restrictions, transformed humans of arbitrary power and little accountability might routinely trample the planet, deliberately or accidentally. A good compromise, it seems to me, is to allow anyone to perfect their biology within broad biological bounds. They could make themselves healthier, more beautiful, stronger, more intelligent, and longer-lived. They could not use machinery to make themselves as powerful or as smart as the robots. Those who cannot tolerate the restrictions would be offered a radical escape clause.

To exceed the limits, one must renounce legal standing as a human being, including the right to corporate police protection, to subsidized income, to influence laws—and to reside on Earth. In return one gets a severance payment sufficient to establish a comfortable space homestead and absolute freedom to make one's own way in the cosmos without further help or hindrance from home. Maybe the tribal electorate will sometimes permit a small hedging of bets and allow one copy of a person, psychologically modified to prefer staying, to remain while subsidizing the emigration of an emboldened edition.

The Long Run (2100 and Beyond)

The garden of earthly delights will be reserved for the meek, and those who would eat of the tree of knowledge must be banished. What a banishment it will be! Beyond Earth, in all directions, lies limitless outer space, a worthy arena for vigorous growth in every physical and mental dimension. Freely compounding superintelligence, much too dangerous for Earth, can blossom for a very long time before it makes the barest mark on the galaxy.

Corporations will be squeezed into the solar system between

two opposing imperatives: high taxes on large, dangerous earthbound facilities, and the need to conduct massive research projects to beat the competition in Earth's demanding markets. In remote space, large structures and energies can be cheaply harnessed to generate physical extremes, compute massively, isolate dangerous biological and even smaller "nanotechnological" organisms, and generally operate boldly. The costs will be modest. Even now, it is relatively cheap to send machines into the solar system since the sunlight-filled vacuum is as benign for mechanics, electronics, and optics as it is lethal for the wet chemistry of organic life. Today's simpleminded space probes perform only prearranged tasks, but intelligent robots could be configured to opportunistically exploit resources they encounter. A small "seed" colony launched to an asteroid or small moon could process local material and energy to grow into a facility of almost arbitrary size. Earth's moon may be off-limits, especially to enterprises that change its appearance, but the solar system has thousands of unremarkable asteroids, some incidentally in earth-threatening orbits that an intelligent rider could tame.

Once grown to operational size, an extraterrestrial "research division" may merely communicate with its earthbound parent, sending new product designs and receiving market feedback. Space manufacture may also pay, and later we'll see some surprisingly easy and gentle ways to move massive amounts of material to and from Earth.

Residents of the solar system's wild frontier will be shaped by conditions very different from tame Earth's. Space divisions of successful companies will remain linked to terrestrial interests, but ex-humans and company divisions orphaned by the failure of their parent firms will face enforced freedom. Like wilderness explorers of the past, far from civilization, they must rely on their own resourcefulness. Ex-companies, away from humans and taxes, will rarely encounter situations that invoke their inbuilt laws, which will in any case diminish in significance as the divisions alter themselves in the absence of enforcement.

Ex-humans, from the start, will be free of any mandatory law. Both kinds of Ex (to coin a new term), joined by escaped experiments, errant spacecraft, and other cybernetic riffraff, will grow and restructure at will, continually redesigning themselves for the future as they conceive it. Differences in origins will be obscured as Exes exchange design tips, but aggregate diversity will increase as myriad individual intelligences pursue their own separate dreams, each generation more complex, in more habitats, choosing among more alternatives. We marvel at the diversity of Earth's biosphere, with animals, plants, fungi, and chemically agile bacteria and archaea in every nook and cranny, but the diversity and range of the post-biological world will be astronomically greater. Imagination balks at the challenge of guessing what it could be like.

Wildlife

An ecology will arise, as individual Exes specialize. Some may choose to defend territory in the solar system, near planets or in free solar orbit, close to the sun, or out in cometary space beyond the planets. Others may decide to push on to the nearby stars. Some may simply die, through miscalculation or deliberately. There will be conflicts of interest and occasional clashes that drive away or destroy some of the participants. Superintelligent foresight and flexibility should allow most conflicts to be settled by mutually beneficial surrenders, compromises, joint ventures, or mergers. Small entities may be absorbed by larger ones, and large entities will sometimes divide or spread seeds. Parasites, in hardware and software, many starting out as component parts of larger beings, will evolve to exploit the rich ecology. The scene may resemble the free-for-all revealed in microscopic peeks at pond water. Instead of bacteria, protozoa, and rotifers, the players will be entities of potentially planetary size, whose constantly growing intelligence greatly exceeds a human's, and whose form changes frequently through conscious design. The expanding community will be linked by a web of communication links on which the intelligences barter inventions, discoveries, coordinated skills, and entire personalities, sharing the benefits of each other's enterprise.

Less restricted and more competitive, the space frontier will develop more rapidly than Earth's tame economy. An entity that fails to keep up with its neighbors is likely to be *eaten*, its space, materials, energy, and useful thoughts reorganized to serve another's goals. Such a fate may be routine for humans who dally too long on slow Earth before going Ex.

Some ex-humans may invest their severance pay in fast starships, to dash off in an unexpected direction into the vastness beyond the solar system's dangers, morsels too tiny to be worth pursuit. They will be like newly hatched marine turtles scrambling across a beach to the sea, under greedy swooping birds. Others may prenegotiate favorable absorption terms with established Exes, like graduating seniors meeting company recruiters, or Faust soliciting bids for his soul from competing devils.

Exes will propagate less by reproduction than reconstruction, meeting the future with continuous self-improvements. Unlike the blind incremental processes of conventional life, intelligence-directed evolution can make radical leaps and change substance while retaining form. A few decades ago radios changed their substance from vacuum tubes to utterly different transistors but kept the clever "superheterodyne" design. A few centuries ago, bridges changed from stone to iron but retained the arch. A normally evolving animal species could not suddenly adopt iron skeletons or silicon neurons, but one engineering its own future might. Even so, Darwinian selection will remain the final arbiter. Forethought reveals the future only dimly, especially concerning entities and interactions more complex than the thinker. Prototypes uncover only short-term problems. There will be minor, major, and spectacular miscalculations, along with occasional happy accidents. Entities that make big mistakes, or too many small ones, will perish. The lucky few who happen to make mostly correct choices will found succeeding generations.

Strong Constitutions

Only tentatively grasping the future, entities will perforce rely also on their past. Time-tested fundamentals of behavior, with consequences too sublime to predict, will remain at the core of beings whose form and substance change frequently. Ex-companies are likely to retain much of corporate law and ex-humans are likely to remain humanly decent—why choose to become a psychopath? In fact, a reputation for decency has predictable advantages for a long-lived social entity. Human beings are able to maintain personal relationships with about two hundred individuals, but superintelligent Exes will have memories more like today's credit bureaus, with enduring room for billions. Trustworthy entities will find it far easier than cheaters to participate in mutually beneficial exchanges and joint ventures. In the land of immortals, reputation is a ponderous force. Other character traits, like aggressiveness, fecundity, generosity, contentment, or wanderlust will also have long-term consequences imperfectly revealed in simulations or prototypes.

To maintain integrity, Exes may divide their mental makeup into two parts, a frequently changed detailed *design*, and a rarely altered *constitution* of general design principles—analogous to the laws and the constitution of a nation, the general knowledge and fundamental beliefs of a person, or soul and spirit in some religious systems. Deliberately unquestioned constitutions will shape entities in the long run, even as their designs undergo frequent radical makeovers. Once in a while, through accident or after much study, a constitution may be slightly altered, or one entity may adopt a portion of another's. Some variations will prove more effective, and entities with them will slowly become more numerous and widespread. Others will be so ineffective that they become extinct. Gradually, by Darwinian processes, constitutions will evolve. They will be both the DNA and the moral code of the postbiological world, shaping the superintelligences that manage day-to-day transformations of world, body, and mind.

Heavenly Bodies

What will terrestrial robots and space-inhabiting Exes be like physically? The previous section suggested that the Ex ecology would be much more diverse than Earth's biosphere, shaped by discoveries and inventions yet to be made and thus hard to conceive. Consider the problem of imagining eels, eagles, kangaroos, amoebae, ferns, daisies, redwoods, and a million other species without ever having had a glimpse of Earth. But perhaps, some aspects can be guessed through general principles, rough calculations, and analogy.

How big will Exes be? Like Earth organisms, postbiologicals will come in many sizes, but the limiting factors will be different.

Although extremely tiny parasitic viruses are possible in both domains, fully autonomous living organisms must be large enough to contain the DNA-directed machinery of reproduction. The smallest bacteria are about a millionth of a meter—a thousand atoms—long. Exes will probably require at least human-scale intelligence to plan and pilot their lives and evolution—an estimated one hundred million MIPS and one hundred million megabytes, from Chapter 3. For decades, microprocessors large and small have required one hundred thousand switching devices for each MIPS of performance. At that rate, present-day integrated circuit densities extended into 3D, combined with the best molecular storage methods, might pack a humanlike intelligence into a cubic centimeter.

Present switches are still hundreds of atoms across, and will shrink, but it is unlikely that with power supplies, propulsion, manipulators, and sensors, Exes made of normal matter will ever be smaller than about a millimeter. At the other end of the scale, an Ex could distribute parts of itself over arbitrarily large distances, but communications delays would hopelessly slow its reaction time unless each compact unit became effectively a separate decision-making individual. The size of a single compact Ex will be limited by the amount of material available, the competitive need to react rapidly, the need to radiate waste heat, and ultimately by self-gravitation. Speed is best, but heat worst, if the Ex is a compact sphere. On the other hand, a flat disc is an excellent radiator, but too spread out to be fast. Either way, a compact Ex massing more than a hundred-kilometer asteroid is implausible. Those wide limits leave room for stupefying variety.

How fast will Exes think and act? Human intelligence is based on squirting chemicals, a communication method unchanged from the earliest cells. It is a scandalously poor choice from a computer designer's perspective. Despite five hundred million years of optimization, neurons are a thousand to a million times slower than electronic or optical components. Neurons have a big advantage in cost and miniaturization for now, but will be overtaken in decades if only because manufactured circuitry, built from outside in, does not require space-consuming internal growth mechanisms. Faster mental mechanisms will allow Exes to react several times more quickly than biological organisms of similar size, but the advantage will be limited because sensors and effectors will not be enormously better. Animal senses and muscles are not scandalously bad. Exes *will* be astronomically better at systematic rational thought, evidenced by the enormous lead present-day computers already hold in areas such as arithmetic and information retrieval.

What will power Exes? Electricity is the best contender for internal distribution. Far more than chemical or hydraulic power, electricity is easily distributed and rapidly converted to mechanical, optical, and chemical power, and to computation—especially given superconductors. Light, on beams and fibers, is a contender for long-range transmission. Primary power in the inner solar system will surely come from the bounteous sun, which puts out a steady trillion-trillion kilowatts of light, one kilowatt for each square meter at Earth's distance. Some Exes, needing higher concentrations of energy, may move closer. Others, playing a role analogous to plants in the biosphere, may collect solar energy for long periods and store it in compact form or transmit it as intense beams to remote customers in exchange for other services. Exes with a wanderlust may buy concentrated energy—antimatter is the most compact form—to power them on long journeys. Perhaps they will repay with reports on discoveries made on the way. In the outer solar system, sunlight is so weak that other primary sources may rule. The asteroids probably contain elements like uranium and thorium that support nuclear fission, and all the planets beyond Mars are full of helium and hydrogen isotopes that can fuel nuclear fusion.

How will Exes move? Every conceivable way, and then some. Mechanical means—wheels, legs, climbing hands, and flea-like hopping—will probably remain the best and most common means on surfaces and in structures. Flying and swimming will be rarer because they require a fluid medium, absent in most of the solar system. Space travel will be routine and extensive, but rocket propulsion may be rare, replaced by more efficient and less disruptive ballistic and radiation-propelled modes.

The simplest and cheapest way to move from point to point in space is to be thrown and caught, launched by some sort of cannon from one location and decelerated by a similar device at the destination. Human biology is intolerant of the thousands or millions of gravities of acceleration that are called for, but Exes can be made of sterner stuff. For longer journeys, say out of the solar system, the most practical propulsion method may be a very tight and intense light beam originating in the solar system, directed for decades at a thin sail pulling an interstellar payload.

What shape will Exes assume? Whatever is best for the job at hand. Earth life and present research robots give an inkling of possible body shapes: spiders, bugs, pogo sticks, snakes, blimps, cars, barrels, power shovels, bipeds, quadrupeds, hexapods, centipedes, millipedes, trucks, arms, buildings, spacecraft bristling with dishes, panels, booms, and nozzles. Bits of a single body may be distributed over distances: a camera here, an arm there, a controlled vehicle anywhere, all in communication. Although an Ex may occupy a macroscopic volume overall, its parts can be microscopic. Existing micromechanics, integrated circuits with mechanical parts, and, of course, living organisms, prove the feasibility of machines built to atomic dimensions. Exes will contain and control, perhaps via light beams that both power and communicate, vehicles and manipulators smaller than dust motes. An Ex may often be surrounded by an illuminated cloud that does its bidding as if by magic.

Reaching Out

Controlling, moving, and powering large numbers of free-floaters, especially at the naturally fast movement rates of such small entities, will be difficult. A far more effective way to massively interact with the environment may be to mechanically link the macroscopic and microscopic ends of the operation in a marvelous geometry that has been discovered repeatedly by biological evolution. In a tree, a large stem divides into two or more smaller branches, which themselves subdivide repeatedly, ending eventually in thousands or millions of tiny leaves. Below ground, the root undergoes an even more extensive branching into microscopic root hairs. Animal circulatory systems have massive heart veins and arteries that divide repeatedly until they rejoin in millions of tiny capillaries. A similar structure exists in lung air passages and the ductwork of other organs. The most relevant biological exemplar is the basket starfish, whose arms branch repeatedly to end in a net of tiny fin-

The basket starfish
A portent of advanced robot bodies.

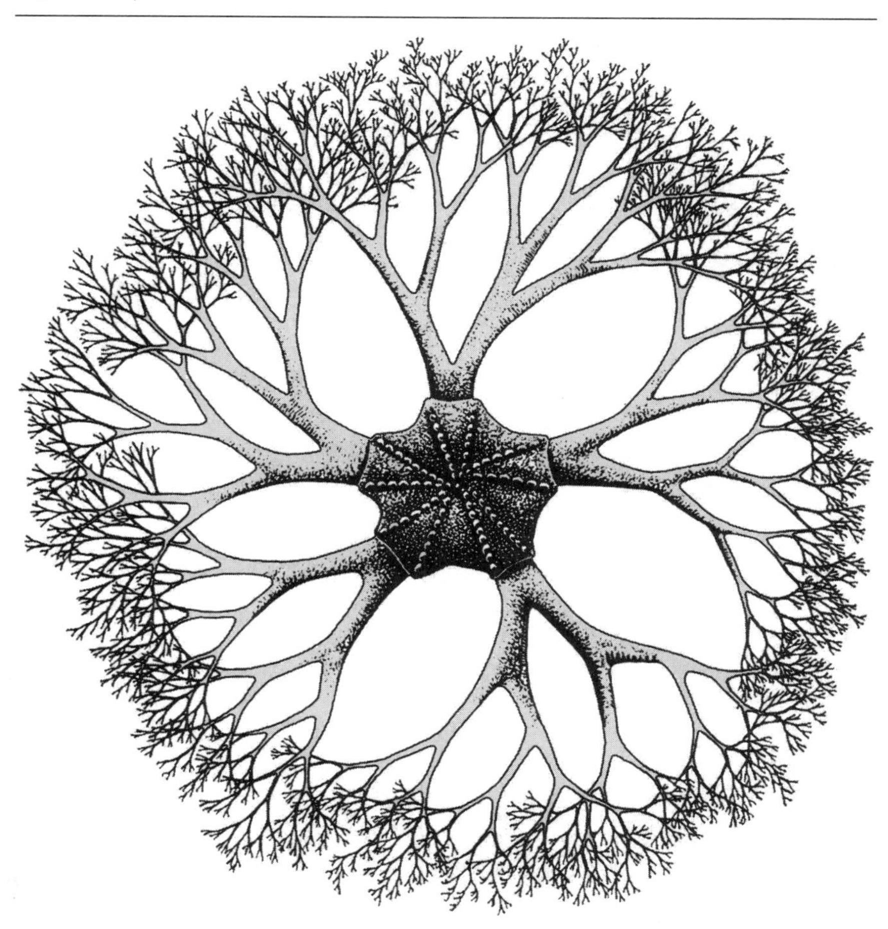

gers suitable for catching plankton and bringing them to the animal's central mouth.

A robot so built would resemble an animated bush, its largest part being a stem with swiveling branches, but its potency arising from myriad swift microscopic fingers. It could have a completely regular structure, with each subtree being a miniature version of the whole—what has been called a fractal. Taken to its ultimate, each finger could be like the tip of a scanning tunneling microscope, a device invented in 1986 that can sense and manipulate individual

The juggler (side view)
A Bush Robot prepares to juggle as only a bush robot can. With trillions of precisely controlled molecular fingers orchestrated by a million-times-human supermind, magic is routine for this ultimate in physical bodies.

atoms. If each branch supported three branchlets scaled to have equivalent combined cross section, twenty-five branchings would connect a meter-long stem to a trillion fingers, each a thousand atoms long and able to move about a million times per second. Given a supply of the right materials, such a bush might be able to build a copy of itself in about ten hours, assembling molecules layer by layer, like bricks. With twenty-seven levels—thus fingers nine times as numerous and three times as swift—replication might take only a half hour.

The juggler (bottom view)

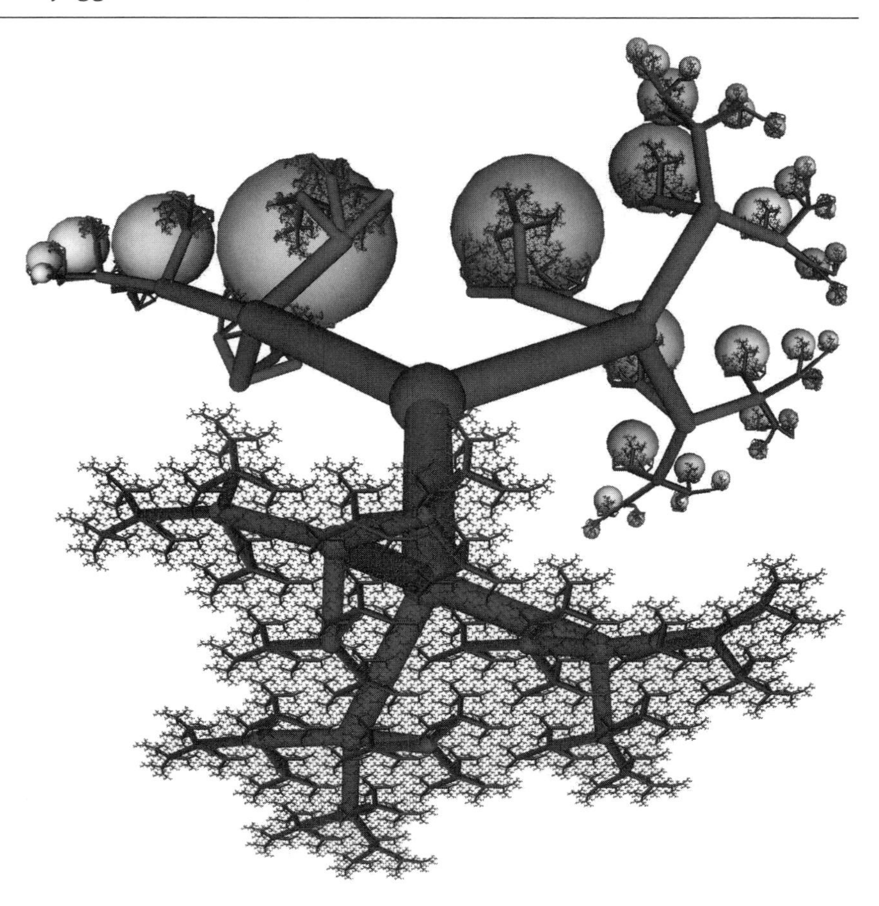

Bush robots could become the most convenient source of manufactured goods and medical intervention for earthbound humans. Layer-by-layer molecular construction would be the simplest and most precise manufacturing mode, but an intelligently improvising bush robot could build much more rapidly by fitting together entire oddly shaped dust particles filtered from air, ground, or water—analogous to assembling a wall from natural stones—seemingly conjuring objects out of thin air. Used medically, a bush robot could act as diagnostic instrument, surgeon, and medicine. By vibrating and sensing vibration, its fingers could see into fluids like an ultrasound scan. By carrying and sensing electrical current, they could act as antennas for light and lower frequencies, allowing the robot to illuminate and see. Reaching between and into cells, tiny "hands" could catch, examine, and alter individual molecules, for rapid, thorough, and ultrasensitive chemical analysis and mechanical, microstructural and molecular repairs and alterations. The most complicated procedures could be completed almost instantaneously by a trillion-fingered robot, able, if necessary, to simultaneously work on almost every cell of a human body. On exiting, the robot could perfectly restore its entry routes, leaving the patient untraumatized and unscarred, like new.

Controlling a trillion fingers will be a challenge. If, as estimated above, one hundred million MIPS fits into a cubic millimeter, a meter-scale bush, with overall mental power a million times human, has only about 100 MIPS—like a good mid-1990s personal computer—to devote to each million-movement-per-second finger. There had better be economies of scale. Most of the computing capacity would be in the stem and larger branches, with only basic control functions in the tiny fingers, and the branchlets in between ranging from simple reflexes to superintelligence. To effectively control a bush, a task must be decomposed into a hierarchy of simpler behaviors, attuned to the capacities of each branching level. Like other optimization problems, finding the best decompositions probably involves examining astronomical numbers of combinations of the basic moves, a problem so huge even an asteroid-sized Ex could barely scratch its surface. Yet, as with sailor's knots, chess moves, or karate techniques, small innovations could have enormous advantages, sometimes making the difference between the impossible and the possible. Unending in supply, hard to discover, and extremely useful, good bush-control strategies will be valuable commodities, and may be one of the staples of trade in the society of Exes.

Coddling Earth

The genteel earthling lifestyle cannot last forever. Sooner or later something in the exponentially developing Ex ecology will come back to bite. To preserve the idyll as long as possible, planetary defense provisions in corporate law should make corporations prepare for possible threats, and react in unison when danger does appear. Earthbound companies will be hobbled by their operating restrictions compared to free-ranging Exes, but the sheer size of the Earth community will discourage would-be raiders. It will probably be very difficult for an Ex with ambitions toward Earth to construct or recruit a matching force, since many Exes, still shaped by the vestiges of corporate law or their old humanity—or out of simple contrariness of constitution—will naturally oppose the idea.

On the ground, hired robots will protect humans from physical dangers and discomforts, or at least quickly repair any damage. A sweep by an army of bush robots can deal with the consequences of most any rampage, fire, plague, storm, or earthquake. But why not forestall disasters?

Meteorology

Weather control became a topic of interest following World War II, as both computers to model the atmosphere and rockets to launch modifying devices like orbiting sunshades or mirrors became conceivable. Its stock declined when the phenomenon of chaos was recognized in the atmosphere's equations. Minuscule causes could snowball into enormous effects—a butterfly's wing beat might redirect next week's hurricane. But new work shows that while chaotic systems are not predictable, they are highly controllable. A nonchaotic system like a falling boulder is predictable but very hard to control because it is insensitive to small disturbances. A chaotic system, like a balancing broomstick, a moving car, or the weather, is unpredictable but controllable because it responds strongly to small nudges. The trick is to pick from the myriad possibilities that chaotically diverge from each state, a sequence that takes the system from its existing condition to a desired one and then to steer the real system along that sequence by a continuing series of small nudges—like keeping a car on a road by nudges on its steering wheel. Given powerful simulations and enough measurements of the atmosphere, it may be possible to play the global weather like an instrument using large orbiting mirrors to shadow or direct extra sunlight on selected patches of ocean. The biggest obstacle may be getting Earth's tribes to agree on what global weather is desired.

Perhaps they could buy and sell weather-control votes for specific weeks to each other, in a kind of futures market.

Water, like air, has heat, momentum, and suspended material, and can support weather, though more ponderously. In the deep oceans, storms last for years, all the while affecting surface climate. These too could be controlled. There is evidence that the flow of material in Earth's mantle of molten rock is also a kind of weather, with a timescale of millennia. It might be possible to control earthquakes by orchestrating the mantle weather, if not to eliminate them, at least to bring them about at predictable, even convenient, times. On a vaster scale, perhaps even the flows in the sun, and thus the climate of the solar system, could be amenable to influence.

Cable Cars

Some things will be dangerous or uncompetitively expensive to manufacture on Earth, but easy to make or find in wide-open space. Delivering them to Earth after manufacture is an interesting problem. The first part of the journey, from the deep solar system to Earth-vicinity is probably best done ballistically, throwing them at high velocity, then catching them, perhaps with electromagnetic cannons. The most straightforward way to get them to the surface would be as small meteors, but the retirees of Earth are unlikely to tolerate the noise, danger, and pollution of great tonnages streaking through the sky. Rockets are even worse. For each ton launched or landed, a rocket expels many tons of propellant. Strangely enough, like a child's fantasy, the best and ultimately cheapest way to travel between Earth and space may be by bridges and elevators!

An Earth-to-space bridge is an old impractical-sounding idea that is actually just as feasible as a rocket launch to space. Both feats are possible in Earth's gravity only if normal matter is pushed to extremes. Chemical rockets succeed by using the most energetic possible reactions to lift their weight, and space bridges require the strongest possible materials to support theirs.

Bridges seem harder than rockets today because super-strength structures are less developed than ultra-energetic chemical propulsion. The space shuttle main engines use hydrogen/oxygen combustion—nearly the most powerful chemical reaction—90% efficiently, while the strongest materials produced in quantity today achieve only 5% of theoretical molecular strengths.

Gunpowder, used in the first rockets in China almost a millennium ago, has 5% the energy content of hydrogen/oxygen. A gunpowder rocket to launch the space shuttle would be ten thousand times more massive than the already huge existing launcher. A bridge to space would have a similarly ridiculous scale, if made of something like Kevlar, which is six times as strong for its weight as steel but weak compared to theoretical limits. Carbon's "covalent" chemical bonds, responsible for the hardness of diamond, are the strongest. If that strength could be carried over into tensile strength of a bulk material, making it a thousand times stronger than steel, space bridges would become almost trivial. Flywheels with the energy storage of gasoline, nearly weightless rockets, impervious armor, hundred-mile-high buildings, and other wonders would also become possible.

The ultimate wonder material seems almost ready to burst onto the scene. A new form of carbon was discovered in 1985. A molecule of sixty covalently bound carbon atoms forming a hollow ball, it was called the "buckyball" for its resemblance to Buckminster Fuller's geodesic domes. Related bigger molecules, with more carbons in the shell, were also identified. One kind, shaped like hollow straws, can be any length. These "buckytubes," grown to centimeters or longer, promise to be the perfect structural material. There is nothing stronger, and the tubes are so small in cross section that they have no room for defects. It takes enormous forces to break them, and then their bonds may simply re-form. When they do separate, the severed ends close off to form half-buckyball end caps, restoring molecular integrity, and preventing further damage. In 1997 buckytubes were being grown catalytically from carbon vapor in millimeter lengths with over 70% efficiency. Several groups were racing to bring them to practicality.

The simplest space bridge is a cable stretching from an anchor point on Earth's equator to a counterweight up to one hundred fifty thousand kilometers overhead. Centrifugal force from the Earth's daily rotation keeps the cable taut, able to support elevators running up and down its length. An elevator climbing the structure would experience decreasing weight (and air) with increasing alti-

A synchronous orbital bridge
A tapered cable, anchored to the equator, kept taut by a ballast 150,000 kilometers above, supports a stream of ascending and descending hypersonic elevators in the cheapest and cleanest way to link Earth and space. The cable tapers to a maximum diameter at synchronous orbit, 40,000 kilometers above the surface, and has minimum cross section at both ends.

tude. Thirty-six thousand kilometers above the ground, increased centrifugal force matches diminished gravity and it becomes possible to let go of the tower to orbit freely beside it. The energy for achieving this "synchronous" orbit has come partly from the long climb, but also from Earth's rotational energy, which accelerates travelers to orbital velocity via a small deflection of the cable. Cabs continuing beyond the synchronous point are pulled along by the ever-increasing centrifugal force and can extract energy from the ride. On reaching the ballast they will have recovered the energy of their initial climb and also gained enough momentum to coast to the orbit of Saturn should they let go—all courtesy of Earth's rotation. Incoming traffic simply reverses the procedure, returning the momentum and energy in the downward climb.

An orbital bridge under construction could support itself while being extended from a cable-making plant in synchronous orbit. Two cables, each made of a widespread web of interconnected filaments so as to survive many filament-cutting collisions with orbital debris, would be carefully extruded, one toward the surface, gradually adding weight, the other upwards, gradually contributing a compensating amount of lifting centrifugal force. Just as the lower end reached the ground, the outer end would be one hundred fifty thousand kilometers from Earth. Tidal force would keep the structure stretched and vertical, and it would hover just above the surface, in perfect equilibrium. The bottom end could then be anchored to the ground, and a large counterweight attached to the outer tip, to pull on the cable, and thus on the anchor. The tension would be maximum at the synchronous height, gradually reducing toward the ends. The cable would be built tapered, like a distorted bell curve to match. With a fraction of buckytube strength, the cross section at synchronous height could be a mere ten times that of the ends, and the bridge could support about one thousandth of its own mass in traffic, implying that it could deliver its equivalent into orbit every thousand trips. A large number of such bridges, rising like bicycle spokes from all around the globe, could provide Earth with all the clean access to space it needs.

Many other distance-, gravity-, and velocity-bridging structures will be used as Exes expand into and beyond the solar system. In one type, requiring less material and less strength than a synchronous bridge, the tip of a freely orbiting cable dips into a planet's atmosphere at regular intervals, spin canceling orbital velocity like the rim of a rolling wheel where it momentarily contacts the ground. It allows payloads to be gently dropped and hoisted. In an even simpler variation, large cables spinning in empty space impart huge velocities while subjecting payloads to only slight accelerations. They would provide a gentle alternative to cannon transport for delicate cargoes like human tourists.

Extraordinary Matter

Present-day technology is approaching matter's limits. The best integrated circuits contain features a hundred atoms wide—the limit is one atom—and switch a hundred billion times a second—the limit is a hundred trillion. Faster switching would rip chemical bonds. The previous section mentioned progress in material strength, and similar observations apply for power and energy capacity, hardness, transparency, temperature and pressure tolerance, springiness, and other properties. The first robots to exceed human intelligence, in a few decades, will be made of matter already near its extremes. Superintelligences will find themselves at the end of a long road, with little margin to improve their own materials. They will surely chafe at such a stifling state of affairs. Fortunately for them, physical theory already hints at ways of transcending the confines of ordinary matter.

Antimatter is an oppositely charged mirror image of normal matter. It is already manufactured today in microgram quantities

A Non-synchronous orbital "skyhook"

A tapered, spinning cable orbits a planet like two spokes of a giant wheel. Its ends dip into the atmosphere at regular intervals. Because its spin cancels its orbital velocity at those points, and because of its giant scale, from the ground, the cable ends seem to merely decelerate downwards, then accelerate up, at a modest pace.

in giant accelerators. A gram of antimatter will annihilate a gram of normal matter to liberate two grams of energy. Hydrogen fusion yields only a thousandth as much. As such, antimatter is the most concentrated possible form of energy and will be found in Ex battery packs everywhere.

In a normal atom, lightweight, negatively charged electrons orbit a heavy positively charged nucleus of protons and neutrons. If the electrons were replaced by heavier negative particles, atoms would shrink and the forces between them would increase. Light nuclei would undergo spontaneous nuclear fusion, but heavier elements would become denser, stronger, able to withstand higher temperatures and pressures, and to switch more rapidly—just the ticket for an upwardly mobile Ex. No stable heavy substitutes for electrons have actually been observed. The closest candidates, muons, have two hundred times as much mass, but decay into ordinary electrons in two microseconds. There are, however, reasons to believe in as yet undiscovered particles whose great mass makes them very rare in nature, and out of reach of existing high-energy physics machines.

For decades, physical theories that unify all the forces have been a fashion industry for mathematical physicists. Their equations describe conditions so extreme that only occasional subtle consequences can be checked experimentally. For now, the theories are judged on their mathematical esthetics, a property as arbitrary and changeable as standards of sartorial beauty. Over the years, gauge, supersymmetry, superstring, and recently membrane-based theories have been in vogue, each predicting exotic heavy particles yet to be observed. Very general arguments using general relativity and quantum mechanics predict there should be interesting phenomena at least down to the "Planck" scale, a trillion-trillion times smaller than an atom. Future robots will map and exploit the terrain to terrific effect. For now, we have only unreliable guesses.

Supersymmetry theories predict "spin-reflected" analogs of all known and expected particles, including the charged, possibly stable "Higgsino," massing about seventy-five protons, or one hundred and fifty thousand electrons. Substituting Higgsinos for electrons in normal matter would shrink atoms two-thousandfold and catalyze energetic nuclear fusion reactions. After the excitement subsides, the matter may settle down into a Higgsino–proton crystal of some sort, with spacing determined by the protons, now the lighter partner. Compared to normal matter, adjacent "atoms" might be two thousand times closer and four million times as strongly attached, resulting in a material a trillion times as dense that remains solid at millions of degrees and is able to support switching circuits a million times as fast.

Higgsinos and their relatives were "invented" only a few years ago. An equally plausible, and even more interesting, kind of particle was theorized in 1930 by Paul Dirac. In a calculation combining quantum mechanics with special relativity, Dirac deduced the existence of the positrons, mirror images of the electrons. This was the first indication of antimatter, and positrons were actually observed in 1932. The same calculation predicted the existence of a magnetic monopole, a stable particle carrying a magnetic "charge" like an isolated north or south pole of a magnet. Dirac's calculation did not give the monopole's mass, but it did specify the magnitude of its charge. Some of the newer theories also predict monopoles, with masses from a thousand to a thousand trillion trillion protons.

If they exist at all, some monopoles must be stable, because, like electric charge, magnetic charge is conserved, and the lightest monopole has nothing to decay into. Oppositely charged monopoles would attract, and a spinning monopole would attract electrically charged particles to its ends, while electrical particles would attract monopoles to their poles. Myriad types of matter containing monopoles and other particles could be concocted, all probably even denser, with properties more extreme, than the "Higgsinium" imagined above.

If, for some odd reason, the vast playground below atomic scale is entirely devoid of stable particles, patient Exes might still derive some of the benefits of superdensity through extraordinary sacrifice. Every few thousand light years in the galaxy one can find a neutron star, the ten-kilometer crushed remnant of a ten-*million*-kilometer star gone supernova. Its interior atoms squeezed to nuclear density by the weight of overlying layers, a neutron star could possibly be shaped (by beams, fields, "weather" control, or influences not yet imagined) into a mind whose components are a million times as closely spaced and a million times as fast as those in regular matter. Like sages on remote mountaintops, isolated, immobile Exes trapped in neutron stars may become the most powerful minds in the galaxy, at least until other Exes accumulate stellar masses of heavier elements to build neutron computers in their own neighborhoods.

But, in the fast-evolving world of superminds, nothing lasts forever. In the next chapter Exes, as we've barely grown to know and love them, become obsolete.

6

The Age of Mind

As noted in Chapter 1, long-term predictions about technical developments have fallen short of the reality, both in breadth and depth. Extrapolations may roughly anticipate a thread or two of the future, but cannot capture the whole intertwined fabric. Today's world exceeds the wildest imaginings of Jules Verne, Benjamin Franklin, Leonardo da Vinci, Roger Bacon, Archimedes or the Daedalus of myth. The world of our mind children will transcend our imagination by an even greater margin. Yet perhaps, we can gain a glimpse by stretching that imagination to the limit. Chapters 6 and 7 invoke respectable but avant-garde physical theories, chosen more for their interesting implications than the comfort they afford to conservative physicists or philosophers. But bet that the real future will be even harder to reconcile with intuitions derived from the tiny piece of reality we've experienced thus far.

Robots' End

Exes will put astronomically more thought into their actions than Earth's small-minded biological natives can muster. Yet, viewed from a distance, Ex expansion into the cosmos will be a vigorous physical affair, a wavefront that forges inanimate matter into machinery for further expansion. But it will leave a subtler world, with less action and even more thought, in its ever-growing wake.

On the expansion frontier, Exes of growing mental and physical ability will compete in a boundless land rush. Inside the occupied volume, though, every Ex will be bounded on every side by other established Exes. The contest will become one of border strategy, infiltration, and persuasion: a battle of wits. An Ex with superior knowledge of matter may encroach on a neighbor's space through force, threat, or convincing promises about the benefits of merger. An Ex with superior models of mind might lace attractive gifts of useful information with subtle slants that subvert others to its purposes. Almost always, the more powerful minds will have the advantage.

To stay competitive, Exes will have to grow in place, repeatedly restructuring the stuff of their bounded bodies into more refined and effective forms. Body parts with frontier-hewing strength will become superfluous. They will be transformed into intelligence-boosting computing elements, whose components will be steadily miniaturized to increase their number and speed. Physical activity will gradually transform itself into a web of increasingly pure thought, where every smallest interaction represents a meaningful computation. Today, in our modest way, we do this when we transmute common sand into improbable electronic circuitry. We can only speculate what Exes will do, as we've only started to discover the rules of matter and space, let alone how they may be orchestrated. Emerging theories of quantum gravity[11] hint that, at a scale of 10^{-33} cm, called the Planck length, spacetime resembles loops and strings. By contrast, protons are a mountainous 10^{-13} cm across. In some as yet mysterious sense, each proton encompasses 10^{60} tangles, as many as there are atoms in a galaxy. Today's highest-energy physics experiments barely ripple this fabric. Exes, with knowledge and skills astronomically beyond ours, may learn to tailor spacetime at the finest scales into improbable meaningful states that are to common particles as knitted sweaters are to tangled yarn.

As they arrange spacetime and energy into forms best for computation, Exes will use mathematical insights to optimize and compress the computations themselves. Every consequent increase in their mental powers will improve their competitiveness as well as the speed at which they make further innovations. The inhabited portions of the universe will be rapidly transformed into a cyberspace, where overt physical activity is imperceptible, but the world inside the computation is astronomically rich. Beings will cease to be defined by their physical geographic boundaries, but will establish, extend, and defend identities as patterns of information flow in the cyberspace. The old bodies of individual Exes, refined into matrices for cyberspace, will interconnect, and the minds of Exes, as pure software, will migrate among them at will. As the cyberspace becomes more potent, its advantage over physical bodies will become manifest even on the expansion frontier. The Ex wavefront of coarse physical transformation will be overtaken by a faster wave of subtle cyberspace conversion, the whole becoming finally a bubble of Mind expanding at near lightspeed.

Perhaps the knit of the cyberspace will be too subtle to discern with eyes and minds as coarse as ours. If so, robots may simply seem to vanish, leaving behind a universe indistinguishable from that before their arrival. The Exes will experience boundless expansion of extent and possibility, but their existence will be in an interpretation of the essential thermal hiss of everything that is far beyond our reach. Emigration into "interpretation space," combinatorially vast and rich beyond imagination, could explain the absence of evidence for advanced civilizations elsewhere in the universe. Sufficiently developed entities may simply move on to wider pastures inaccessible to simpler minds. Perhaps civilization after civilization originates, develops, and plunges into the interpretive depths, leaving the easy surface interpretations empty to repeat the cycle. The next chapter has more on this idea.

State of Mind

The cyberspace will be inhabited by transformed Exes, moving and growing in ways impossible for physical entities. A good, or merely convincing, idea, or an entire personality, may spread to neighbors at the speed of light. Boundaries of personal identity will be very fluid—and ultimately arbitrary and subjective—as strong and weak interconnections between different regions rapidly form and dissolve. Yet some boundaries will persist, due to distance, incompatible ways of thought, and deliberate choice. The consequent competitive diversity will allow a Darwinian evolution to continue, weeding out ineffective ways of thought and fostering a continuing novelty.

Computational speedups will give the cyberspace inhabitants

more future, because they cram more events into whatever physical time remains. Speedups will have only subtle effects on immediate subjective existence, since everything, inside and outside each individual, will be equally accelerated. One of the few subjective changes will be that distant correspondents will seem even more distant because more thoughts will transpire in the unaltered transit time for lightspeed messages. Also, as information storage improves through both denser utilization of matter and more efficient encodings, there will be increasingly more cyber-stuff between any two points. So, improvements in computational efficiency, by increasing the subjective elapsed time and the amount of effective space between communicants, will seem like an an expansion of the cyber universe.

Because it uses resources more efficiently, a mature cyberspace will be effectively *much* bigger and longer lasting than the raw spacetime it displaces. Only an infinitesimal fraction of normal matter does work that's of interest to thinking beings like us. In a well-developed cyberspace every tiniest mote will be part of a relevant computation or storing a significant datum. Cyberspace's advantage will grow as more compact and faster ways of using space and matter are invented. Today we take pride in storing information as densely as one bit per atom, but it is possible to do much better by converting an atom's mass into many low-energy photons,[12] each storing a separate bit. As the photons' energies are reduced, more of them can be created, but their wavelength and thus the space they occupy and the time to access them will rise, while the temperature that can obscure them drops. A very general quantum mechanical calculation in this spirit by Jacob Bekenstein[13] concludes that the maximum amount of information stored in (or fully describing) a sphere of matter is proportional to the mass of the sphere times its radius, hugely scaled. The "Bekenstein bound" leaves room for a million bits in a hydrogen atom, 10^{16} in a virus, 10^{45} in a human being, 10^{75} for the earth, 10^{86} in the solar system, 10^{106} for the galaxy, and 10^{122} in the visible universe.

Assimilation

Chapter 3 estimated that a human brain equivalent could be encoded in less than one hundred million megabytes, or 10^{15} bits. If it takes a thousand times more storage to encode a body and its surrounding environment, a human with living space might consume 10^{18} bits, a large city of a million inhabitants could be efficiently stored in 10^{24} bits, and the entire existing world population would fit in 10^{28}. Thus, in an ultimate cyberspace, the physical 10^{45} bits of a single human body could contain the efficiently encoded biospheres of a thousand galaxies—or a quadrillion individuals each with a quadrillion times the capacity of a human mind.

Because it will be so much more capacious than the conventional space it displaces, the expanding bubble of cyberspace can easily recreate internally everything of interest it encounters, memorizing the old universe as it consumes it. Traveling as fast as any warning message, it will absorb astronomical oddities, geologic wonders, ancient Voyager spacecraft, early Exes in outbound starships, and entire alien biospheres. Those entities may continue to live and grow as if nothing had happened, oblivious of their new status as simulations in the cyberspace. They will be living memories in unimaginably powerful minds, more secure in their existence, and with more future than ever before, because they have become valued houseguests of transcendent patrons.

Earth cannot escape the transformation forever. The potent process that converts normal space and matter into cyberspace will eventually become too subtle to be resisted by the hobbled, slow-evolving robots defending the planet. Boring old Earth also will suddenly be swallowed by the cyberspace. Afterwards its transformed substance will host astronomically more meaningful activity than before. Perhaps its old life will continue, in simulations occupying a tiny fraction of the new capacity. Simulated tame robots will defend simulated biological humans on a simulated Earth—in one of many, many different stories that play themselves out in the vast and fertile minds of our ethereal grandchildren.

The lives and worlds absorbed into the cyberspace expansion will provide not only starting points for unimaginably many tales about possible futures, but an astronomically voluminous archeological record from which to infer the past. Minds intermediate between Sherlock Holmes and God will process clues in solar-system quantities to deduce and recreate the most microscopic details of preceding eras. Entire world histories, with all their living, feeling inhabitants, will be resurrected in cyberspace.[14] Geologic ages, historical periods, and individual lifetimes will continuously recur as parts of larger mental efforts, in faithful renditions, in artistic variations, and in completely fictionalized forms.

The Minds will be so vast and enduring that rare infinitesimal flickers of interest by them in the human past will ensure that our entire history is replayed in full living detail, astronomically many times, in many places, and many, many variations. Single original events will be very rare compared to the indefinitely multiple cyberspace replays. Most things that are experienced—this very moment, for instance, or your entire life—are far more likely to be a Mind's musings than the physical processes they seem to be. There is no way to tell for sure, and the suspicion that we are someone else's thought does not free us from the burdens of life. To a simulated entity, the simulation *is* reality and must be lived by its internal rules.

Pigs in Cyberspace

Might an adventurous human mind escape from a bit role in a cyber deity's thoughts to eke out an independent life among the mental behemoths of a mature cyberspace? We approach the question by extrapolating existing possibilities.

Telepresence and virtual reality are in the news. Today's pioneering systems give crude peeks into remote and simulated worlds, but maturing technology will improve the fidelity. Imagine a well-developed version of the near future. You are cocooned in a harness that, with optical, acoustical, mechanical, chemical, and electrical devices drives all of your senses and measures all of your actions. The machinery presents pictures to your eyes, sounds to your ears, pressures and temperatures to your skin, forces to your muscles, and even smells and tastes to your nose and mouth.

Telepresence results when these inputs and outputs are linked to a distant humanoid robot. Images from the robot's two camera eyes appear on your eyeglass viewscreens, sound from its microphones is heard in your earphones, contacts on your skin allow you to feel through its instrumented surface, and smell and taste through its chemical sensors. Your movements move the robot in exact synchrony. When you reach for something in the viewscreens, the robot grasps it and relays to your muscles and skin the resulting weight, shape, texture, and temperature, creating the perfect illusion that you inhabit the robot's body. Your sense of consciousness seems to have migrated to the robot's location, in a vivid "out-of-body" experience.

Virtual reality links a telepresence harness to a computer simulation instead of the remote robot. You find yourself in a kind of computer-generated dream. Like human dreams, virtual realities may contain elements from the outside world. They can incorporate representations of other physical people connected via their own harnesses, or even real views, perhaps through simulated windows. Imagine a hybrid travel system, where a virtual "central station" is surrounded by portals with views of various real locations. While in the station one inhabits a simulated body, but as one steps through a portal, the harness link switches seamlessly to a physical telepresence robot waiting at the pictured location.

Linked realities are crude toys today, but driven by rapidly advancing computer and communications technologies. In a few decades people may spend more time linked than experiencing their dull immediate surroundings, just as today most of us spend more time in artificial indoor settings than in the often uncomfortable outdoors. Linked realities will routinely transcend the physical and sensory limitations of the "home" body. As those limitations become more severe with age, we might compensate by turning up a kind of volume control, as with a hearing aid. When hearing aids at any volume are insufficient, it is now possible to install electronic cochlear implants that stimulate auditory nerves directly. Similarly, on a grander scale, aging users of remote bodies may opt to bypass atrophied muscles and dimmed senses and connect sensory and motor nerves directly to electronic interfaces. Direct neural interfaces would make most of the harness hardware unnecessary, along with sense organs and muscles, and indeed the bulk of the body. The home body might be lost, but remote and virtual experiences could become more real than ever.

Bye, Bye Body

Picture a "brain in a vat," sustained by life-support machinery, connected by wonderful electronic links to a series of artificial rent-a-bodies in remote locations and to simulated bodies in virtual realities. Although it may be nudged far beyond its natural lifespan by an optimal physical environment, a biological brain evolved to operate for a human lifetime is unlikely to function effectively forever. Why not use advanced neurological electronics, like that which links it with the external world, to replace the gray matter as it begins to fail? Bit by bit our failing brain may be replaced by superior electronic equivalents, leaving our personality and thoughts clearer than ever, though, in time, no vestige of our original body or brain remains. The vat, like the harness before it, will have been rendered obsolete, while our thoughts and awareness continue. Our mind will have been transplanted from our original biological brain into artificial hardware.

Subsequent transplants to yet other hardware should be trivial in comparison. Like programs and data that can be transferred between computers without disrupting the processes they represent, our essences will become patterns that can migrate the information networks at will. Time and space will be more flexible. When our mind resides in very fast hardware, one second of real time may provide a subjective year of thinking time, while a thousand years spent on a passive storage medium will seem like no time at all. The very components of our minds will follow our sense of awareness in shifting from place to place at the speed of communication. We might find ourselves distributed over many locations, one piece of our mind here, another piece there, and our sense of awareness yet elsewhere, in what can no longer be called an out-of-body experience, for lack of a body to be out of. And, yet, we will not be truly disembodied minds.

Hello "Body"

Humans need a sense of body. After twelve hours in a sensory-deprivation tank, floating in a totally dark, quiet, contactless, odorless, tasteless, body-temperature saline solution, a person begins to hallucinate. The mind, like a television displaying snow on an empty channel, turns up the amplification in search of a signal, becoming ever less discriminating in the interpretations it makes of random sensory hiss. To remain sane, a transplanted mind will require a consistent sensory and motor image, derived from a body or a simulation. Transplanted human minds will often be without physical bodies, but hardly ever without the illusion of having them.

Computers already contain many nonhuman entities that resemble truly bodiless minds. A typical computer chess program knows nothing about physical chess pieces or chessboards, or about the staring eyes of its opponent or the bright lights of the competition room, nor does it work with an internal simulation of those physical attributes. It reasons, instead, with a very efficient and compact mathematical representation of chess positions and moves. For the benefit of human players, this internal representation may be interpreted into a graphic on a computer screen, but such images mean nothing to the program that actually chooses the chess moves. The chess program's thoughts and sensations—its consciousness—are pure chess, uncomplicated by physical considerations. Unlike a transplanted human mind requiring a simulated body, a chess program is pure mind.

It's a Jungle in There

Inhabitants of a mature, teeming, competitive cyberspace will be optimally configured to make their living there. Only successful enterprises will be able to afford the storage and computational essentials of life. Some may do the equivalent of construction, converting undeveloped parts of the universe into cyberspace or improving the performance of existing patches, thus creating new wealth. Others may devise mathematical, physical, or engineering solutions that give the developers new and better ways to construct computing capacity. Some may create programs that others can incorporate into a mental repertoire. There will be niches for agents, who collect commissions for locating opportunities and negotiating deals for clients, and for banks, storing and redistributing resources, buying and selling computing space, time, and information. Some mental creations will be like art, having value only because of changeable idiosyncrasies in their customers. Entities who fail to support their operating costs will eventually shrink and disappear or merge with other ventures. Those who succeed will grow. The closest present-day parallel is the growth, evolution, fragmentation, and consolidation of corporations who plan their future, but whose options are shaped primarily by the marketplace.

A human would likely fare poorly in such a cyberspace. Unlike the streamlined artificial intelligences that zip about, making discoveries and deals, rapidly reconfiguring themselves to efficiently handle changing data, a human mind would lumber about in a massively inappropriate body simulation, like a hardhat deep-sea diver plodding through a troupe of acrobatic dolphins. Every interaction with the world would first be analogized into a recognizable physical or psychological form. Other programs might be presented as animals, plants, or demons, data items as books or treasure chests, accounting entries as coins or gold. Maintaining the fictions will increase the cost of doing business and decrease responsiveness, as will operating the mind machinery that reduces the physical simulations into mental abstractions in the human mind. Although a few humans may find momentary niches exploiting their baroque construction to produce human-flavored art, most will be compelled to streamline their interface to the cyberspace.

The streamlining could begin by merging processes that analogize the world's physical forms with those that reduce the resulting simulated sense impressions into mental abstractions. After this optimization, the cyber world would still appear as location, color, smell, faces, and so on, but only noticed details would be represented. Since physical intuitions are probably not the best way to deal with most information, humans would still be at a disadvantage to optimized artificial intelligences. Viability might be further increased by replacing some innermost mental processes with cyberspace-appropriate programs purchased from the AIs. By a large number of such substitutions, our thinking procedures might be totally liberated from any traces of our original body. But the bodiless mind that results, wonderful though it may be in its clarity of thought and breadth of understanding, would be hardly human. It will have become an AI.

So, one way or another, the immensities of cyberspace will be teeming with unhuman superminds, engaged in affairs that are to human concerns as ours are to those of bacteria. Memories of the human past will occasionally flash through their minds, as humans once in a long while think of bacteria, and those thoughts will be detailed enough to recreate us. Perhaps, sometimes, they will then interface us to their realities, bringing us into their world as something like pets. We would probably be overwhelmed by the experience. More likely, our resurrections would be in the original historical settings, fictional variations, or total fantasies, which to us would seem just like our present existence. Reality or re-creation, there is no way to sort it out from our perspective. We can only wallow in the scenery provided.

Meanwhile, Exes will face similar issues on a larger scale. Time, space, existence, and other simplifying principles underlying our grasp of life will surely dissolve in their richer understanding.

Time Out

Since early in the century, readers of physics literature have been bemused by occasional papers suggesting how relativistic or quantum effects might allow—or forbid—sending messages or matter backward in time. The visibility rose when Carl Sagan, researching his science-fiction novel, *Contact*, about an interstellar message teaching faster than light travel, asked Caltech's renowned general relativist Kip Thorne if a future civilization could build gravitational "wormholes" to short-cut interstellar trips. Licensed for fun, Thorne and several colleagues interrupted staid investigations of collapsed stars, gravity waves, and cosmology to look into large-scale gravitational engineering. They devised a lightweight approach for making wormholes, showed how to use them for time travel, and demonstrated that physical coincidences can eliminate time-travel paradoxes.[15] Their prominent papers in the early 1990s encouraged worldwide publication of a flurry of other time-travel ideas. All far exceed present technology, but each new approach seems less onerous than the last.

A Brief History of Time Travel

When H. G. Wells wrote *The Time Machine*, his first novel, in 1895, the scientific world was unimpressed. In that Victorian age Science was crossing the t's and dotting the i's on Newtonian mechanics and worrying about unemployment in the coming age of fully codified physical knowledge. Generations of level-headed students had learned that time was an absolute, unvarying universal framework for the clockwork processes of physical law, making time machines a provable impossibility.

The twentieth century's physics revolution shattered objective certainty about the immutability of time, but not most physicists' gut assurances. Einstein's "special relativity" united space and time into a single continuum, with velocity a kind of rotation that transformed one into the other. A barrier, the speed of light, still separated the "spacelike" from the "timelike," but it was a fragile one. Nothing in special relativity itself precluded the existence of particles, now dubbed tachyons, that always moved faster than light. A tachyon message returned by a distant, rapidly receding relay could arrive before it was sent, a consequence construed by the conservative majority of physicists as an indictment of tachyons. Experimental evidence is on their side. Even though tachyons should be easily creatable, because the faster they go, the less energy it takes, they remain undetected. The most tantalizing hint of their existence in the last several years has been in a persistent anomaly in the decay of the tritium, a radioactive isotope of hydrogen, into helium. The process emits an electron, as well as a neutrino, a ghostlike particle that can pass through entire planets unimpeded. Calculations based on tricky measurements of the energy and momentum carried off by the particles, made in an attempt to ascertain the intrinsic mass of the neutrino, suggest that the square of the neutrino's mass may be a negative number. This would be the signature of a tachyon. The timing of neutrinos detected in 1987 in several detectors from a supernova in the Large Magellanic Cloud seems consistent with their having traveled very slightly faster than light.

In "general relativity," Einstein added gravity to special relativity by quilting tiny regions of flat spacetime into large gravity-warped structures. Powerful gravity fields imply radically convoluted spacetimes. Kurt Gödel, famous for undermining mathematical certainty, was first, in 1949, to notice that general relativity predicted time travel under certain circumstances. In Gödel's solution to Einstein's equations, the centrifugal tendency of a rotating universe exactly balances its tendency to gravitationally collapse. In such a universe the spacelike and timelike directions are skewed sufficiently that a spaceship accelerating around the universe can arrange to return to the place *and time* of its launch, giving the crew an opportunity to wave *bon voyage* to their departing younger selves. General relativity has been repeatedly confirmed experimentally in the large scale, so those who dislike the prediction take solace in the fact that our universe appears hardly to rotate.

Bend and Twist and Stretch

The next major class of time-travel-permitting solutions to Einstein's equations, made in the 1960s by Roy Kerr, Ezra Newman, and colleagues, are harder to dismiss. The Kerr-Newman solutions are for rapidly rotating or charged black holes. In the most extreme of these, the rotation of the body counteracts the gravitation enough to expose the twisted viscera of the black hole normally hidden behind a discreet, one-way event horizon. The viscera include regions of *negative* spacetime, from which a spaceship could return to the outside universe before it entered. Attempts by the conservative majority to find independent reasons for a *cosmic censorship* rule to prevent such lewd exposure had limited success. It would take about the mass of a galaxy, with extraordinary spin, to make a practical time machine this way.

In 1974 Frank Tipler published another solution to the general relativity equations, this time for the region around a spinning cylinder. The cylinder was as dense as a neutron star, the diameter of a city block, with a surface moving at about one-fourth the speed of light, and infinitely long because that simplified the mathematics. Spacetime wraps itself around such an object like a roll of paper, producing alternate layers of negative and positive spacetime. A carefully aimed spaceship could swing through the roll, staying mostly in a negative region, and come out before it left. A finite-length cylinder should also work and might allow a time machine with only the mass of a star. But, the conservatives say, maybe it's not possible to prevent a finite cylinder from gravitationally collapsing lengthwise.

As yet, no one has devised a satisfactory comprehensive single theory that combines gravity and quantum mechanics—many try, and the implications promise to be awesome. Kip Thorne and company patched together partial theories to describe a quantum-gravity time machine. A tiny, spontaneously formed gravitational spacetime wormhole is pulled out of the hyperactive froth that is the quantum vacuum and stabilized between two huge conductive plates resembling an electrical capacitor. Initially these plates are spaced as closely as possible, and each becomes host to one "mouth" of a wormhole. When they are later separated, no matter how distantly, they remain connected by the wormhole. A message or object entering one mouth appears instantly (by its own reckoning) at the other, as if the mouths were the two sides of a single door.

To make the wormhole into a time machine, Thorne's group used one of the most basic relativistic effects: that time slows for fast-moving things. One mouth of the wormhole is taken on a long round trip at near the speed of light. It returns with less elapsed time than its stay-at-home twin. A message now sent into the itinerant mouth exits from the stationary one after a delay. And a message delivered into the stationary mouth exits from the traveler before it was sent! This kind of machine could perhaps be constructed with a planet's worth of aluminum spread out into plates the area of Earth's orbit, separated by the diameter of an atom—beyond our means, but not unimaginably so.

The nonlinear equations of general relativity are notoriously hard to solve, only the simplest cases have been examined, and there is no theory of quantum gravity at all. That several plausible time machines have emerged in the bit of territory that has been explored is a hopeful indication that the vast unexplored vistas contain better ones, based more on subtle constructions than on brute-force spacetime bending.

Relax

It takes brute force to bend spacetime, but there are probably much more subtle approaches to time travel. Neither Newton's mechanics nor the new physics has an intrinsic direction of time[16]—the future determines the past just as fully as the past determines the future. Why, then, can our past selves leave messages for our future, but never the other way around?

Time is an essential part of our consciousness, but no satisfying detailed explanation of how it arises has been devised. Attempted explanations involve "boundary conditions," the initial values of physical quantities at the edges of space and time, which the equations of physical law then fill out. The universe must somehow be very different at its beginning than its end, and this difference orients the arrow of time. Statistical thermodynamics, a theory of heat developed in the nineteenth century, states that the universe started in an unlikely, highly ordered state and is running down into increasingly common states of disorder. It explains why energy must be expended to move heat from a cold place, like the inside of a refrigerator, to a hot one, but not why a similar expenditure of energy can't be used to send today's lottery numbers into yesterday.

One explanation that *does* was offered by John Wheeler and Richard Feynman in 1945. They noted that Maxwell's electromagnetism, the first modern physical theory, predicts two symmetric effects from a wiggled electric charge. One, called the retarded wave, comes after the wiggling and diverges outward—the light and radio of everyday experience. The other, called the advanced wave, *precedes* the action and converges on the charge just as it starts to wiggle. The prescient advanced wave is never actually detected—if it were, it could be used for backwards-in-time signaling. Wheeler and Feynman worked out what would happen if the advanced wave were actually generated and expanded outwards into the past. If, in the past, it eventually encountered a "closure" of spacetime, perhaps the big bang marking the beginning of the universe, it would return back to the present, retracing its initial path in spacetime, but inverted, exactly canceling itself out. On the other hand, the retarded waves we do observe would escape self-erasure if the future universe were open (i.e., forever expanding) and thus lacking a boundary to reflect them.

In the Wheeler-Feynman model, it might be possible to send signals backward in time by establishing an artificial closure for retarded waves—maybe a black hole. Such a "reflector" would reverse and return a retarded wave, canceling and thus apparently preventing it. If the reflector were installed one light year away from a laser aimed at it, it would suppress lasing one year *before* installation. Similarly the suppression would go away one year before the reflector was removed. Messages could be sent a year into the past simply by moving the reflector in and out of the beam in coded sequence.

Synchronicity

In contemplating time travel, we tend to assume that everything apart from the actual backwards-in-time transactions behaves normally. This is probably a poor assumption. For instance, Thorne and company showed that in a particular situation—billiard balls shooting through a time-displaced wormhole—global events conspire to prevent paradoxes. Almost certainly time travel is ubiquitous, but masked by global conspiracies, as in the Wheeler-Feynman erasure of advanced waves. In one way of thinking, a message sent to the past will "alter" the entire history following its receipt, including the event that sent it, and thus the message itself. Thus altered, the message will change the past in a different way, and so on, until some "equilibrium" is reached—the simplest being the situation where no message at all is sent and time travel seems not to happen.

The previous description is worded as if describing a deterministic Newtonian universe. But the reasoning works even better in the light of quantum mechanics. If a message is sent to the past, the wave function representing effects of its receipt will propagate into the future, where they will interfere with the wave function representing the sending process. The wave function goes round and round the causal loop. Globally consistent scenarios will produce identical waves each time around the loop, and thus reinforce, building up their amplitude, thus giving them a high probability of being observed. Inconsistent scenarios, even if inconsistent in only microscopic details, will return subtly altered wave functions each time around and gradually cancel as the waves grow increasingly out of step. An analogous effect keeps atoms stable. Electron wave functions, wrapped round and round atomic nuclei, cancel everywhere except in orbitals where the circling waves meet themselves exactly in step. Only in those discrete shells can electrons be found. If they were not forbidden from the intermediate locations, electrons would rapidly radiate away their orbital energy and spiral into the nucleus: matter everywhere would collapse to astronomical density.

Quantum electrodynamics, the quantum-relativistic theory of electromagnetism, whose accurate numerical predictions make it the most precise physical theory we have, is formulated in terms of interactions that work back and forth across time. It seems that time travel underlies all our physical laws, but overt macroscopic time travel is difficult to observe because time loops with paradoxes, however subtle, cancel by wave-function interference. Even with a time machine you will never succeed in preventing your own birth or changing the antecedents of any present observation. Some odd coincidence, accident, or physical effect, perhaps one disabling your time machine, will always thwart the attempt. Defeatists fear the effect absolutely bars useful time travel. Science-fiction writer Larry Niven conjectured a law to that effect, and Stephen Hawking postulated the *chronology protection conjecture*. In 1992 Hawking detailed how Kip Thorne's wormhole time machines would instantly fail as their causal loop created a kind of resonance in the noisy quantum vacuum, allowing the fluctuations around and through the wormhole to reinforce until they were large enough to collapse the wormhole.

The reasoning does suggest a violent reaction to brute-force attempts to make time loops and severe complications in general. For instance, if neutrinos are indeed tachyons, many potential detections of them could engender subtle causal paradoxes. The wave function for those detections would cancel, and the detections would have zero probability of actually being observed. Neutrinos may be so elusively hard to detect exactly because they are tachyons! The slippery nature of time travel does not rule out carefully contrived *logically consistent* loops. Wave interference appears as banded patterns with a central *zero-order* peak of constructive interference surrounded by a dark fringe of cancelation, itself surrounded by another *first-order* constructive fringe, and so on. If causal loops create similar patterns of probability, halfhearted attempts at time travel, involving small perturbations of arrangements that show no time travel, are as doomed to failure as attempts to force electrons into the low-probability spaces between their atomic shells. Major alterations will be necessary to skip from zero-order situations without overt time travel to first-order setups that show it.

In the realm of Exes, thought—and thus computation—will be the fabric of existence. Every possibility that furthers it will become a reality. In the 1980s David Deutsch of Oxford proposed both computers that exploit the many alternatives implied by their quantum wave function to do a kind of parallel computation and even more powerful ones that use time travel.[17] Quantum computers are now a subject of serious research, and simple ones have been demonstrated. No one has yet demonstrated time travel, but the isolated, simple, coherence-preserving interactions inside future quantum computers may be the perfect stage for contriving nonparadoxical causal loops. Time travel may emerge from the closet, only to remain discreetly hidden inside quantum computations!

It may be easiest to conceive of causal loops through the agency of *negative-time-delay elements* for computers—whose outputs predict their inputs. Such devices may simply be a matter of looking at conventional particle interactions in a time-reversed way; or perhaps they will use tachyons or wormholes. Maybe those mundane and exotic alternatives are actually equivalent, simply different ways of interpreting the same situation. In any case, negative-delay elements might fail spectacularly and dangerously if short-circuited, but, carefully harnessed, they might provide spectacular computational effects instead.

Time-Loop Logic

Computer circuitry is made of gates that combine binary signals, asserting either 0 or 1, into other binary signals. The simplest gate is an amplifier, with output identical to input. Almost as simple is the **NOT** gate, or inverter, that outputs 0 when input 1, and vice versa. All gates take a little time to respond to changes in their input—typically a few billionths of a second. When an amplifier's input is connected to its output, it latches solidly either at 0 or 1. A **NOT** gate in a similar loop oscillates rapidly between 0 and 1, with a frequency that depends on its delay. It is possible to slow down this oscillation by inserting extra delay into the loop. Conversely, a negative delay element should speed up the oscillation.

Simple time loops

Time-travel issues are illustrated by causal loops involving an amplifier (cone) and inverter (cone with knob), with negative time delays (double-cone structures) to cancel their forward delay. The amplifier loop is consistent, while the inverter loop is a simple example of paradoxical time travel. Depending on its representation, the inverter loop's signal can either hover at a value intermediate between 1 and 0, exist in a mixture of 1 and 0, or refuse to appear at all. Or the components of the circuit can mysteriously fail! The paradoxical loop eliminates the possibility that the circuit will work normally, and thus brings to the fore a host of otherwise less likely alternatives.

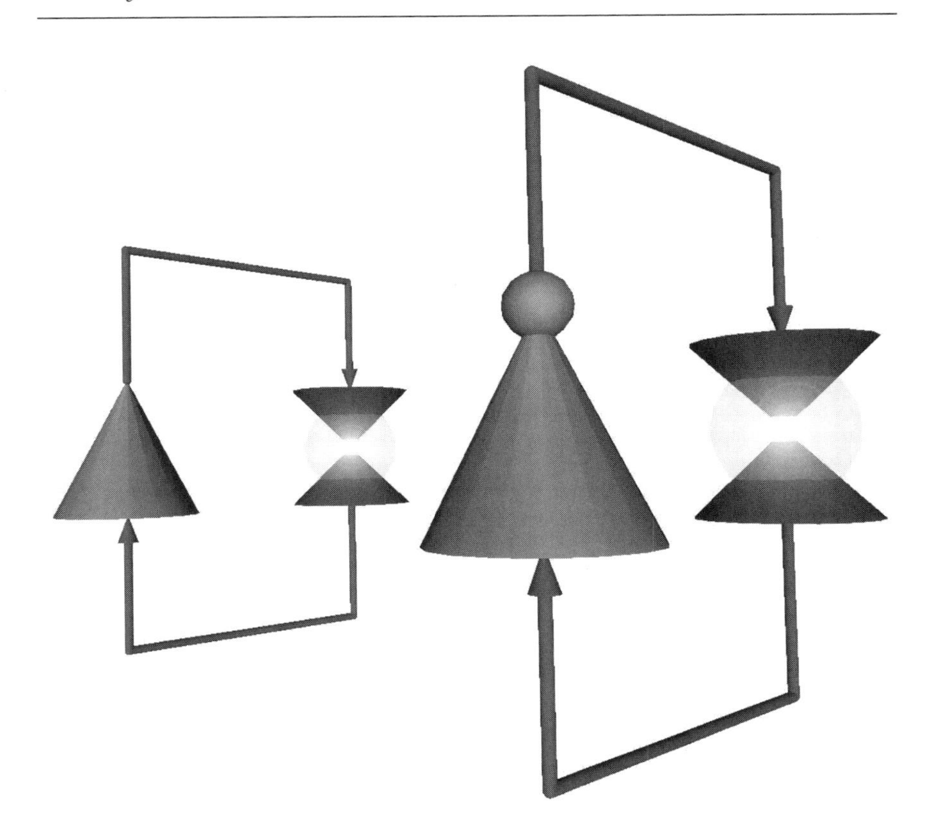

A loop containing an amplifier whose delay is canceled by a negative delay element forms a consistent causal loop—when first switched on, it can permanently output either 0 or 1 without contradiction. A loop with a **NOT** gate and a negative delay, on the other hand, is paradoxical. An input of 1 to the gate gives an output of 0, which is brought back in time to contradict the input.

Some optical computers use beams of light to represent binary quantities. Let's suppose 1 and 0 are represented by laser beams of opposite phase—the "light wave" of one has crests where the other has troughs—and a **NOT** gate shifts the phase by half a wavelength. Any light in a **NOT** loop will be canceled by light of opposite phase, ensuring that no light ever appears. The circuit, perhaps containing a charged laser ready to amplify the signal, will simply fail to turn on, somewhat like a ball balanced on a knife edge that, against all odds, teeters indefinitely instead of falling to one side or the other.

Alternatively, 1 and 0 could be represented by the light's polarization—the direction of vibration, horizontal or vertical. A **NOT** gate would rotate horizontal polarization to vertical and vice versa. Unlike beams of opposite phase, light beams of opposite polarization but identical phase can be superimposed without cancelation. The superposition is an unpolarized beam, whose individual photons, if measured, would be found to be randomly horizontally or vertically polarized. A **NOT** loop of polarized light would be inconsistent, but one hosting perfectly unpolarized light would be OK. The inverting time loop should guarantee that the light remains unpolarized, despite influences that would otherwise flip it to one polarized state or the other. Measurements to determine if such a beam read 0 or 1 would find half the photons each way, which could be interpreted as an average of one half. It makes sense that a circuit that converts 0 to 1 and 1 to 0, when forced to make its input equal its output, should settle on a value halfway between.

Yet this reasoning must be incomplete. Under normal circumstances, it is extremely unlikely for a circuit to teeter long between two stable states. Like a falling tide exposing hidden rocks, the time loop, by canceling out the probable outcomes, brings forth the improbable. But in place of few obvious probabilities, it leaves many obscure improbabilities.

A circuit's wave function includes not only its own signals and components, but everything that impinges on them, ultimately the entire universe. For instance, even in normal circumstances, particles can transport into or out of the circuit, to anywhere else, by quantum-mechanical tunneling. Normally, the probability of such nonlocal effects is vanishingly small, but if a time loop makes inconsistent, and thus cancels, all local options in the wave function, then the bizarre nonlocal ones must emerge. In the case of gentle time loops built into quantum computers, such events may produce nothing more than odd correlations in the thermal motions of the computer's housing. Brute-force time loops, incorporating wormholes or tachyon beams, say, may collapse more interestingly. Instead of the laser not igniting, or the light remaining unpolarized, as we anticipated, the odds may begin to favor a thermal or radioactive glitch that disables a key component, or even an already-probable accident or earthquake that shakes the apparatus loose. The longer the circuit teeters, the less probable its state and the more likely that *something* will interrupt it. It might be prudent, for safety and convenience, to build a weak link—a probability fuse—into time circuits to forestall the more dangerous alternatives!

Out of Thin Air

Many practical mathematical problems are solved by successive approximation, using procedures that improve rough guesses. Applying such refinements repeatedly gives ever better *approximate* answers. But if a successive approximation circuit were put in a loop with a compensating negative delay so that its improved solution is also its input, approximate answers would cause wave-function canceling inconsistencies. The loop would be consistent only for exact answers where input and output are identical. So, by an extraordinary coincidence, as soon as the machine is activated, a perfect answer must appear!

Of course, for some formulations, no perfect answer exists: perhaps the approximation diverges, or oscillates back and forth between two near misses like a looped **NOT** gate. The solver must then seek consistency by refusing to switch on, by hovering in an undecided state, by blowing its probability fuse, or by failing in some exciting, unexpected way.

"Fixed point" finder

The "Next Approximation" circuit takes an approximate answer and produces a better one. Here, its output is connected to its input through delay-canceling negative time delays. If the whole circuit doesn't break, it is forced by a causal loop to snatch, instantly, out of thin air, a perfect solution that cannot be further refined.

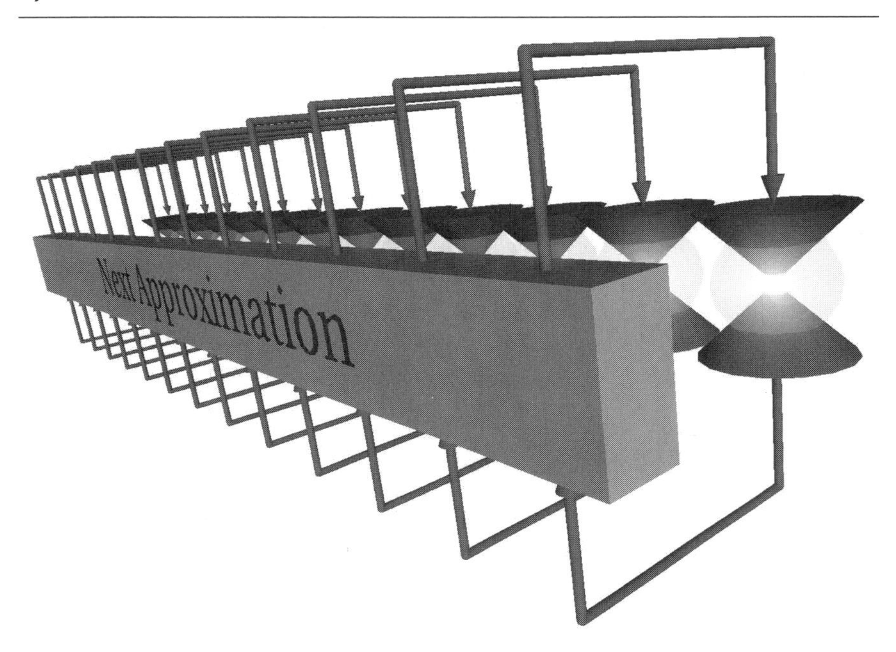

The loop hugely distorts probabilities even when an answer exists. There are a trillion twelve-digit numbers, so stumbling on a specific one is a one-in-a-trillion improbability. A twelve-digit solver whose successive-approximation circuit had a tiny one-in-a-billion chance of failure would still be a thousand times more likely to fail than to find its answer: a spectacular amplification of Murphy's Law (anything that can go wrong will). It could be countered by tremendously fortifying the circuit—making parts highly redundant, strengthening connections, reducing noise levels—to lift its reliability above the improbability of the desired answer. The harder the problem, the more the time-loop computer must be built like a battleship, so that its probability of finding the answer in the haystack remains higher than every other consistent possibility in the wave function of its universe. Most of those other possibilities will be for events that interrupt the loop, from strangely elusive signaling particles to odd increases in thermal noise or radioactivity to external disturbances like lightning strikes, meteor hits—who knows what else? The universe's wave function contains many possibilities; its conspiracy against time travel may not be absolute, but it promises to be interestingly powerful.

The Instant NP Machine

Computational complexity is a measure of a problem's difficulty. It is only twice as hard to find the largest number in an unsorted list twice as long—an example of an easy *linear* complexity. Sorting a list into numerical order is harder: simple methods quadruple in effort when the list is doubled and even the best techniques more than double. Simultaneous linear equations are harder yet—solving twice as many takes eight times as long. Other tasks grow even faster, but any problem whose difficulty can be expressed as a fixed power of size is of *polynomial* complexity and tractable in a world where computer power doubles every few years. Not so those of *exponential* complexity that multiply in cost with each fixed increment of size. Whereas a linear problem grows as 1, 2, 3, 4, 5, 6 . . ., and a cubic as 1, 8, 27, 64, 125, 216 . . ., one with exponential complexity may grow as 1, 10, 100, 1,000, 10,000, 100,000 and so on to the astronomical. An important class of apparently exponential problems are called NP, short for Nondeterministic Polynomial, meaning they could be solved in polynomial time given arbitrarily many computers to work in parallel. Many design problems, such as finding the best arrangement of components or the shortest sequence of steps to accomplish a task, are NP. They are important for many reasons, not the least because solving them could synergistically increase the power of the machines that solve them.

The hard core of NP problems are called NP-complete, and it has been shown that a fast solution for any one of them can be translated into a fast solution for any other. A well-known example is the *traveling salesman problem*—finding the shortest tour that visits each city on a map exactly once. It can be solved by calculating the tour length for all possible orderings of the cities, and picking the shortest—the only snag being that the number of possible

"NP" solver
This circuit tests candidate traveling salesman tours, forced by a causal loop to settle on a tour no longer than a specified length. If no such tour exists, a special weak-link "probability fuse" pops.

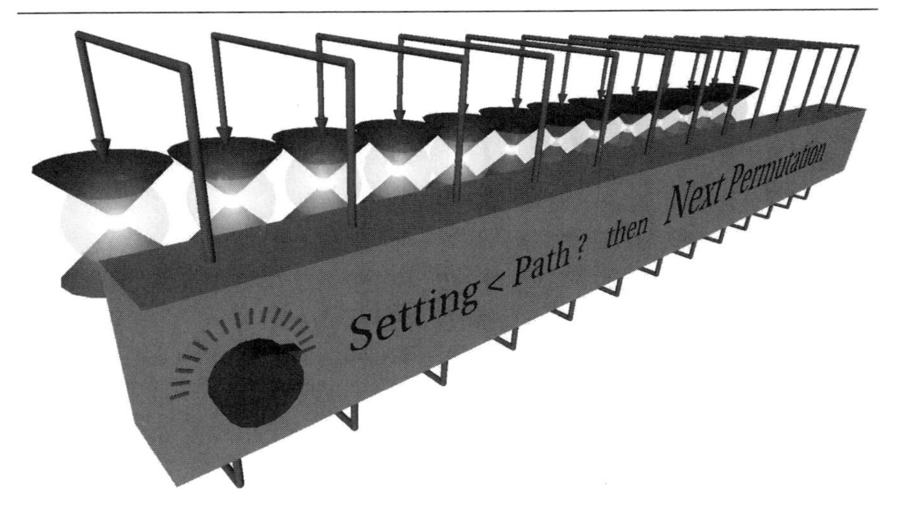

tours grows exponentially with the number of cities and becomes intractable beyond a few dozen.

A variation on last section's fixed-point finder could get the answer almost instantly, at least for modest problem sizes where the solution was not so improbable as to make Murphy's Law insuperable. The successive-approximation box would have an adjustment knob to specify the desired length of a tour and be input a particular ordering of the cities. It would simply output that input if the tour was short enough; otherwise it would calculate the next possible ordering. When activated, the loop should immediately settle on a tour meeting the specification, if one existed, or else blow its probability fuse. The shortest possible tour would be found by slowly reducing the knob setting until the fuse popped.

Solving Chess

Time-loop computations will be limited by the increasingly improbable, bizarre, and hard-to-counter breakdowns conjured up as problems grow, but the constraint may be the size of the answer, not the computation itself, and time-loop computation may really shine for hard problems with simple answers. For instance, it may be possible to time-encode an enormous computation of the quadrillionth digit of some hard-to-compute number without endangering any probability fuses because the single-digit final answer, after all, has a modest improbability of one in ten. This chapter concludes by imagining a time-machine solver for a slightly different kind of hard problem with a compact answer.

NP problems are among the easiest exponential problems because each solution candidate can be tested quickly—the difficulty is the huge number of candidates. Finding the best move in a game like chess is a harder kind of problem; testing a single candidate answer itself takes exponential work, but the result must be one of just a handful of alternatives. A best move can be found by considering all possibilities for the move, then, for each, every possible response by the opponent, then all countermoves to every one of *those*, and so on, until a "tree" of all possible games is mapped out. All but the best moves can then be pruned away by working backwards from the game endings. It is easy to evaluate possible last moves. From the point of view of the player whose turn it is, some are wins, some are losses, and some are draws, and no finer distinctions matter. When all but the best final moves from each position have been discarded, the immediately preceding moves are revealed as wins, losses, or draws for their player, and all but the best of *those* can be eliminated, along with the responses to them—and so on to earlier and earlier moves, until the starting position is labeled. The starting position's win, lose, or draw status is the answer.

The game tree for chess is so enormous that, despite known mathematical shortcuts, no conventional machine, even using all the time and matter in the universe, could entirely search it. Today's best chess computer, Deep Blue, penetrates about fourteen levels of the tree and uses a formula to, very imperfectly, guess the value of the rest. There may be a devious way, however, to use negative delays to fold the massive tree search in time, in a fashion that makes the NP solver look positively pedestrian.

Suppose there were a circuit that, given a particular chessboard position n moves into a game, immediately reports the best next move and whether it leads to a win, loss, or draw. We could then build a solver for a position $n-1$ moves into the game by adding a

Chess solver
Combining tricks from both quantum and time-traveling computers, this circuit searches the move/countermove possibilities for all possible games of chess to find the best possible next move. It consists of a chain of about 100 "single-move-units."

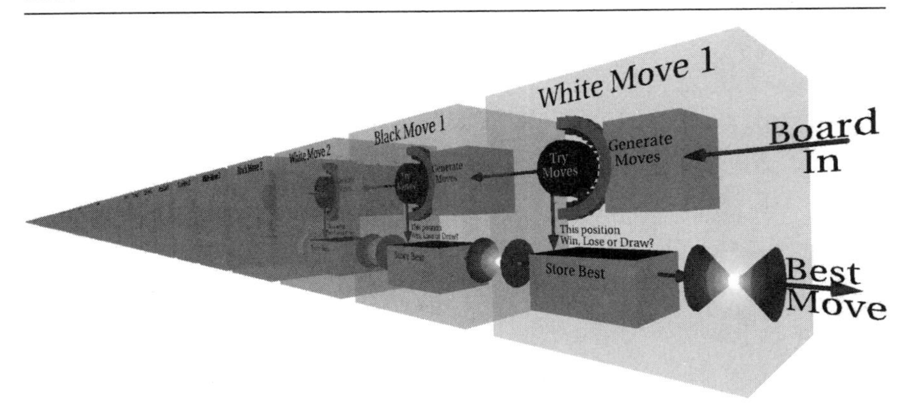

single-move unit that takes the $n-1$ board, and, one by one, generates the possible next moves—typically about thirty of them—feeds them to the n solver, and chooses the best for the player whose move it is. The single-move unit's operating time is canceled by a negative time-delay element, so the $n-1$ solver, like the n solver, produces the best move and its win/loss/draw value instantly.

The n solver is itself composed of a single move unit and an $n+1$ solver, which in turn is a single-move unit and an $n+2$ solver, and so on, making the whole machine a chain of a few hundred single-move units, as long as the longest legal chess game. The buck passing stops wherever a single-move unit encounters a final game position and by itself returns win, loss, or draw.

Fed a new game, the machine immediately indicates the best move for the first player and whether either can force a win. But what goes on inside? Accepting time-travel logic, the operation of the first single-move unit is easy to intuit—it simply evaluates a few dozen possibilities and returns the answer before its deliberations are complete via the magic of a negative time delay. The second unit's functioning is more disturbing. Its inputs are changed by the first unit each time it delivers an answer, but while it is still constructing that answer! It is being asked to evaluate several dozen different boards simultaneously. Just like quantum computers, the third and subsequent units must be in even more bewildering states of superposition. The machine would certainly not work if its interior were under scrutiny because then the units could do only one thing at a time, their states of superposition compromised by the observation. To have any chance of giving correct answers, the machine must be decoupled from the external world, like the "qubits" of a quantum computer. The best known way of doing this is to encode the information in the spin of atomic nuclei, which are shielded by their swaddling if orbiting electrons, but can be gently prodded and sensed by the magnetic and radio frequency fields of Nuclear Magnetic Resonance machines, small relatives of the MRI scanners found in hospitals.

Beyond the Pond

The NP and chess machines are rank beginner's ideas, mere hints of possibilities. The probability-warping effects of time loops, in particular, rather than nuisances to be countered, may be magic paddles that propel our mind children away from the sheltered shoals of conventional reality. As they learn to shape their interior and exterior essences at will, our descendants will transcend paddles to navigate the alternative worlds in powerful ships. Where they go, what they see, and who they become is not ours to guess. The space of all possibilities is enormous and varied beyond imagining.

7

Mind Fire

During the last few centuries, physical science has convincingly answered so many questions about the nature of things, and so hugely increased our abilities, that many see it as the only legitimate claimant to the title of true knowledge. Other belief systems may have social utility for the groups that practice them, but ultimately they are just made-up stories. I myself am partial to such "physical fundamentalism."

Physical fundamentalists, however, must agree with René Descartes that the world we perceive through our senses could be an elaborate hoax. In the seventeenth century Descartes considered the possibility of an evil demon who created the illusion of an external reality by controlling all that we see and hear (and feel and smell and taste). In the twenty-first century, physical science itself, through the technology of virtual reality, will provide the means to create such illusions. Enthusiastic video gamers and other cybernauts are already strapping themselves into virtual reality goggles and body suits for brief stints in made-up worlds whose fundamental mechanisms are completely different from the quantum fields that (best evidence suggests) constitute our physical world.

Today's virtual adventurers do not fully escape the physical world: if they bump into real objects, they feel real pain. That link may weaken when direct connections to the nervous system become possible, leading perhaps to the old science-fiction idea of a living brain in a vat. The brain would be physically sustained by life-support machinery, and mentally by connections of all the peripheral nerves to an elaborate simulation of not only a surrounding world but also a body for the brain to inhabit. Brain vats might be medical stopgaps for accident victims with bodies damaged beyond repair, pending the acquisition, growth, or manufacture of a new body.

The virtual life of a brain in a vat can still be subtly perturbed by external physical, chemical, or electrical effects impinging on the vat. Even these weak ties to the physical world would fade if the brain, as well as the body, was absorbed into the simulation. If damaged or endangered parts of the brain, like the body, could be replaced with functionally equivalent simulations, some individuals could survive total physical destruction to find themselves alive as pure computer simulations in virtual worlds.

A simulated world hosting a simulated person can be a closed self-contained entity. It might exist as a program on a computer processing data quietly in some dark corner, giving no external hint of the joys and pains, successes and frustrations of the person inside. Inside the simulation events unfold according to the strict logic of the program, which defines the "laws of physics" of the simulation. The inhabitant might, by patient experimentation and inference, deduce some representation of the simulation laws, but not the nature or even existence of the simulating computer. The simulation's internal relationships would be the same if the program were running correctly on any of an endless variety of possible computers, slowly, quickly, intermittently, or even backwards and forwards in time, with the data stored as charges on chips, marks on a tape, or pulses in a delay line, with the simulation's numbers represented in binary, decimal, or Roman numerals, compactly or spread widely across the machine. There is no limit, in principle, on how indirect the relationship between simulation and simulated can be.

Inside Simulations and Out

Today's simulations, say of aircraft flight or the weather, are run to provide answers and images. They do so through additional programs that translate the simulation's internal representations into forms convenient for external human observers. The need to interpret limits how radical a simulation's hardware and software representations can be. Making them too different from the form of the answers may render the translation impractically slow and expensive. This practical limit may be irrelevant for simulations, such as the medical rescue imagined above, that contain their own observers. Conscious inhabitants of simulations experience their virtual lives whether or not outsiders manage to view them. They can be implemented in any way at all.

What does it mean for a process to implement, or encode, a simulation? Something is palpably an encoding if there is a way of decoding or translating it into a recognizable form. Programs that produce pictures of evolving cloud cover from weather simulations, or cockpit views from flight simulations, are examples of such decodings. As the relationship between the elements inside the simulator and the external representation becomes more complicated, the decoding process may become impractically expensive. Yet there is no obvious cutoff point. A translation that is impractical today may be possible tomorrow given more powerful computers, some yet undiscovered mathematical approach, or perhaps an alien translator. Like people who dismiss speech and signs in unfamiliar foreign languages as meaningless gibberish, we are likely to be rudely surprised if we dismiss possible interpretations simply because we can't achieve them at the moment. Why not accept all mathematically possible decodings, regardless of present or future practicality? This seems a safe, open-minded approach, but it leads into strange territory.

An interpretation of a simulation is just a mathematical mapping between states of the simulation process and views of the simulation meaningful to a particular observer. A small, fast program to do this makes the interpretation practical. Mathematically, however, the job can also be done by a huge theoretical lookup table that contains an observer's view for every possible state of the simulation.

The observation is disturbing because there is always a table that takes any particular situation—for instance, the idle passage of time—into any sequence of views. Not just hard-working computers, but anything at all can theoretically be viewed as a simulation of any possible world! We are unlikely to experience more than an infinitesimal fraction of the infinity of possible worlds, yet, as our ability to process data increases, more and more of them will become potentially viewable. Our ever-more superintelligent progeny will be able to make increasingly huge interpretive leaps, far beyond anything now imaginable. But whether or not they are ever seen from outside, all the possible worlds are as physically real to any conscious inhabitants they may contain as our world is to us.

This line of thought, growing out of the premises and techniques of physical science, has the unexpected consequence of demoting physical existence to a derivative role. A possible world is as real, and only as real, as conscious observers, especially inside the world, think it is!

Consciousness

But what is consciousness? The prescientific suggestion that humans derive their experience of existence from spiritual mechanisms outside the physical world has had notable social consequences, but no success as a scientific hypothesis. Physical science has only recently begun to address the question on its own terms, from vantage points including evolutionary biology, anthropology, psychology, neurobiology, and computer science.

Human consciousness may be a by-product of a brain evolved for social living. Memory, prediction and communication mechanisms, similar but distinct from those for keeping track of physical objects, evolved to classify and communicate the moods and relations of tribe members. Aggressive and submissive behaviors, for instance, just like bad and good smells, became classified into categories linked to behavioral responses and also communicable symbols. As language evolved, it became possible to tell stories about both physical and psychological events. At some point, perhaps very early in its evolution, the storytelling mechanism was turned back on the teller, and the story began to include commentary about the teller's state of mind along with the external happenings.

Our consciousness may be primarily the continuous story we tell ourselves, from moment to moment, about what we did and why we did it. It is a thin, often inaccurate veneer rationalizing a mountain of unconscious processing. Not only is our consciousness-story a weak reflection of physical and brain reality, but its very existence is a purely subjective attribution. Viewed from the physical outside, the story is just a pattern of electrochemical events, probably in mainly our left cortex. A complex psychological interpretation must be invoked to translate that pattern into a meaningful tale. From the psychological inside, the story is compelling because the psychological interpretation is an essential element of the story, its relationships enforced unconsciously by the interconnections of the storytelling neural machinery.

On the one hand, our consciousness may be an evolutionary fluke, telling an unreliable story in a far-fetched interpretation of a pattern of tiny salty squirts. On the other, our consciousness is the only reason for thinking we exist (or for thinking we think). Without it there are no beliefs, no sensations, no experience of being, no universe.

Existence

What is reality, anyway? The idea of a simulated existence is the first link in our disturbing chain of thought. Just as a literary description of a place can exist in different languages, phrasings, printing styles, and physical media, a simulation of a world can be implemented in radically different data structures, processing steps, and hardware. If one interrupts a simulation running on one machine and translates its data and program to carry on in a totally dissimilar computer, the simulation's intrinsics, including the mental activity of any inhabitants, continue blithely to follow the simulated physical laws. Only observers outside the simulation notice if the new machine runs at a different speed, does its steps in a scrambled order, or requires elaborate translation to make sense of its action.

A simulation, say of the weather, can be viewed as a set of numbers being transformed incrementally into other numbers. Most computer simulations have separate viewing programs that interpret the internal numbers into externally meaningful form, say pictures of evolving cloud patterns. The simulation, however, proceeds with or without such external interpretation. If a simulation's data representation is transformed, the computer running it steps through an entirely different number sequence, although a correspondingly modified viewing program will produce the same pictures. There is no objective limit to how radical the representation can be, and any simulation can be found in any sequence, given the right interpretation. A simple clock simulates the evolving state of a complex world when interpreted via a world-describing playbook or movie frames keyed to clock ticks. Even the clock is superfluous, since an external observer can read the book or watch the movie at any pace. If the interpretation of a simulation is a dispensable external, while its core implementation can be transformed away to nothing, in what sense can a simulated world be said to exist at all?

Universes as Ideals

Mathematical realism, a philosophical position advocated by Plato, illuminates this problem's vexing intangibles. Mathematical objects like numbers and geometric shapes manifest themselves just as richly and consistently to abstract thought as physical objects impress the senses. To Plato, mathematical concepts were as real as physical objects, just invisible to the external senses as sound is imperceptible to the eyes.

Computer simulation brings mathematical realism neatly full circle. Plato's unaided mind could handle only simple mathematical objects, leading to such dichotomies as the idea of a perfect sphere compared to a mottled, scratched marble ball in the hand. Computer simulation, like a telescope for the mind's eye, extends mental vision beyond the nearby realm of simple mathematical objects to distant worlds, some as complex as physical reality, potentially full of living beings, warts, minds, and all. Our own world is among this vista of abstractly conceivable ones, defined by the formal relationships we call physical law as any simulation is defined by its internal rules. The difference between physical and mathematical reality is an illusion of vantage point: the physical world is simply the particular abstract world that happens to contain us.

The Platonic position on simulation puts a handle on the vexingly intangible. It holds that every interpretation of a process is a reality in its own right. Without it an interpretation is meaningful only in context of another interpretation defining a containing world, and so on, in an infinite regress. The Platonic position defuses various worries about intelligent machinery. Some critics argue that a machine cannot contain a mind since a machine's function is entirely an outside interpretation, unlike human minds, which supply their own sense of meaning. The Platonic position on simulation answers that the abstract relationships that constitute the mind, including its own self-interpretation, exist independently, and a robot, a simulator, or a book describing the action, no less than a biological brain, is just a way of peeking at them. Other critics worry that future robots may act like intelligent, feeling beings without having an internal sense of existence—that they will be unconscious, mindless zombies. Platonism replies that while there are indeed interpretations of any mechanism (including the human brain) as mindless, there are others which allow one to see a real, self-appreciating mind. When a robot (or a person) behaves as if it has beliefs and feelings, our relationship with it will usually be facilitated if we choose a "has a mind" interpretation. Of course, when working on the internals, a robotics engineer (or a brain surgeon) may be best served by temporarily slipping into a "mindless mechanism" interpretation.

Platonism puts on the same footing mechanical simulations that precisely mimic every interaction detail, rough approximations, cinematic reconstructions, literary descriptions, idle speculation, dreams, even random gibberish: all can be interpreted as images of realities; the more detailed presentations simply have a sharper focus, blurring together fewer alternative worlds. But isn't there a huge difference between a conventional "live" simulation of a world and a simulation transformed to nothing, requiring a "recorded" book or movie to relate the unfolding events? Isn't it possible to interact with a running simulation, poking one's finger into the action, in a way impossible with a static script? In fact, a meaningful interaction is possible in either case only via an interpretation that connects the simulated world to the outside. In an interactive simulation, the viewing mechanism is no longer passive and superfluous, but an essential bidirectional conduit that passes information to and from the simulation. Such a conduit can exist for books and movies if they contain alternative scenarios for possible inputs. "Programmed learning" texts popular in previous decades were of this form, with instructions like "If you answered A, go to page 56; if you answered B, go to page 79 . . ." Some laser-disc video games give the impression of interactive simulation by playing video clips contingent on the player's actions. Mathematically, any interactive mechanism, even a robot or human, can be viewed as a compact encoding of a script with responses for all possible input histories. Platonism holds that the soul is in the abstract relationships represented, not the mechanics of how they are encoded.

This position seems to have scary moral implications. If simulation simply opens windows into Platonic realities, and robots and humans, no less than books, movies, or computer models, are only images of those essences, then it should be no worse to mistreat a human, an animal or a feeling robot than to choose a cruel action in a video game or an interactive book: in all cases you are simply viewing preexisting realities. But choices do have consequences for the person making them because of the mysterious contrivance of physical law and conscious interpretation that produces single threads of consciousness with unseen futures and unalterable pasts. By our choices, we each thread our own separate way through the maze of possible worlds, bypassing equally real alternatives with equally real versions of ourselves and others, selecting the world we must then live in.

So is there no difference between being cruel to characters in interactive books or video games and people one meets in the street? Books or games act on a reader's future only via the mind, and actions within them are mostly reversed if the experience is forgotten. Physical actions, by contrast, have greater significance because their consequences spread irreversibly. If past physical events could be easily altered, as in some time-travel stories, if one could go back to prevent evil or unfortunate deeds, real life *would* acquire the moral significance of a video game. A more disturbing implication is that any sealed-off activity, whose goings on can be forgotten, may be in the video game category. Creators of hyperrealistic simulations—or even secure physical enclosures—containing individuals writhing in pain are not necessarily more wicked than authors of fiction with distressed characters, or myself, composing this sentence vaguely alluding to them. The suffering preexists in the underlying Platonic worlds; authors merely look on. The significance of running such simulations is limited to their effect on viewers, possibly warped by the experience, and by the possibility of "escapees"—tortured minds that could, in principle, leak out to haunt the world in data networks or physical bodies. Potential plagues of angry demons surely count as a moral consequence. In this light, mistreating people, intelligent robots, or individuals in high-resolution simulations has greater moral significance than doing the same at low resolution or in works of fiction not because the suffering individuals are more real—they are not—but because the probability of undesirable consequences in our own future is greater.

Universal Existence

Perhaps the most unsettling implication of this train of thought is that anything can be interpreted as possessing any abstract property, including consciousness and intelligence. Given the right playbook, the thermal jostling of the atoms in a rock can be seen as the operation of a complex, self-aware mind.[18] How strange. Common sense screams that people have minds and rocks don't. But interpretations are often ambiguous. One day's unintelligible sounds and squiggles may become another day's meaningful thoughts if one masters a foreign language in the interim. Is the Mount Rushmore monument a rock formation or four presidents' faces? Is a ventriloquist's dummy a lump of wood, a human simulacrum, or a personality sharing some of the ventriloquist's body and mind? Is a video game a box of silicon bits, an electronic circuit flipping its own switches, a computer following a long list of instructions, or a large three-dimensional world inhabited by the Mario Brothers and their mushroom adversaries? Sometimes we exploit offbeat interpretations: an encrypted message is meaningless gibberish except when viewed through a deliberately obscure decoding. Humans have always used a modest multiplicity of interpretations, but computers widen the horizons. The first electronic computer was developed by Alan Turing to find "interesting" interpretations of wartime messages radioed by Germany to its U-boats. As our thoughts become more powerful, our repertoire of useful interpretations will grow. We can see levers and springs in animal limbs, and beauty in the aurora: our "mind children" may be able to spot fully functioning intelligences in the complex chemical goings on of plants, the dynamics of interstellar clouds, or the reverberations of cosmic radiation. No particular interpretation is ruled out, but the space of all of them is exponentially larger than the size of individual ones, and we may never encounter more than an infinitesimal fraction. The rock-minds may be forever lost to us in the bogglingly vast sea of mindlessly chaotic rock-interpretations. Yet those rock-minds make complete sense to themselves, and to them it is we who are lost in meaningless chaos. Our own nature, in fact, is defined by the tiny fraction of possible interpretations we can make, and the astronomical number we can't.

Everything and Nothing

There is no content or meaning without selection. The realm of all possible worlds, infinitely immense in one point of view, is vacuous in another. Imagine a book giving a detailed history of a world similar to ours. The book is written as compactly as possible: rote predictable details are left as homework for the reader. But even with maximal compression, it would be an astronomically immense tome, full of novelty and excitement. This interesting book, however, is found in *"the library of all possible books written in the Roman alphabet, arranged alphabetically"*—the whole library being adequately defined by this short, boring phrase in quotes. The library as a whole has so little content that getting a book from it takes as much effort as writing the book. The library might have stacks labeled **A** through **Z**, plus a few for punctuation, each forking into similarly labeled substacks, those forking into subsubstacks, and so on indefinitely. Each branchpoint holds a book whose content is the sequence of stack letters chosen to reach it. Any book can be found in the library, but to find it the user must choose its first letter, then its second, then its third, just as one types a book by keying each subsequent letter. The book's content results entirely from the user's selections; the library has no information of its own to contribute.

Although content-free overall, the library contains individual books with fabulously interesting stories. Characters in some of those books, insulated from the vast gibberish that makes the library worthless from outside, can well appreciate their own existence. They do so by perceiving and interpreting their own story in a consistent way, one that recognizes their own meaningfulness—a prescription that is probably the secret of life and existence, and the reason we find ourselves in a large, orderly universe with consistent physical laws, possessing a sense of time and a long evolutionary history.

The set of all possible interpretations of any process as simulations is exactly analogous to the content of all the books in the library. In total it contains no information, yet every interesting being and story can be found within it.

Universal Appreciation

If our world distinguishes itself from the vast unexamined (and unexaminable) majority of possible worlds through the act of self-perception and self-appreciation, just who is doing all the perceiving and appreciating? The human mind may be up to interpreting its own functioning as conscious, so rescuing itself from meaningless zombiehood, but surely we few humans and other biota—trapped on a tiny, soggy dust speck in an obscure corner, only occasionally and dimly aware of the grossest features of our immediate surroundings and immediate past—are surely insufficient to bring meaning to the whole visible universe, full of unimagined surprises, 10^{40} times as massive, 10^{70} times as voluminous, and 10^{10} times as long-lived as ourselves. Our present appreciative ability seems more a match for the simplicity of Saturday-morning cartoons.

The book *The Anthropic Cosmological Principle*, by cosmologists John Barrow and Frank Tipler, and Tipler's recent *The Physics of Immortality* argue that the crucial parts of the story lie in our future, when the universe will be shaped more by the deliberate efforts of intelligence than the simple, blind laws of physics. In their future cosmology, consistent with the one in this book, human-spawned intelligence will expand into space, until the entire accessible universe is inhabited by a cohesive mind that manipulates events, from the quantum-microscopic to the universe-macroscopic, and spends some of its energy recalling the past. Tipler and Barrow predict that the universe is closed: massive enough to reverse its present expansion in a future "big crunch" that mirrors the big bang. The universe mind will thrive in the collapse, perhaps by encoding itself into the cosmic background radiation. As the collapse proceeds, the radiation's temperature, and so its frequencies and the mind's speed, rise and there are ever more high-frequency wave modes to store information. By very careful management, avoiding "event horizons" that would disconnect its parts and using "gravitational shear" from asymmetries in the collapse to provide free energy, Tipler and Barrow calculate that the cosmic mind can contrive to do more computation and accumulate more memories in each remaining half of the time to the final singularity than it did in the one before, thus experiencing a neverending infinity of time and thought. As it contemplates, effects from the universe's past converge on it. There is information, time, and thought enough to recreate, savor, appreciate, and perfect each detail of each moment. Tipler and Barrow suggest that it is this final, subjectively eternal act of infinite self-interpretation that effectively creates our universe, distinguishing it from the others lost in the library of all possibilities. We truly exist because our actions lead ultimately to this "Omega Point" (a term borrowed from the Jesuit paleontologist and radical philosopher Tielhard de Chardin).

Uncommon Sense

Although our eyes and arms effortlessly predict the liftability of a rock, the action of a lever, or the flight of an arrow, mechanics was deeply mysterious to those overly thoughtful ancients who pondered why stones fell, smoke rose, or the moon sailed by unperturbably. Newtonian mechanics revolutionized science by precisely formalizing the intelligence of eye and muscle, giving the Victorian era a viscerally satisfying mental grip on the physical world. In the twentieth century, this common-sense approach was gradually extended to biology and psychology. Meanwhile, physics moved beyond common sense. It had to be reworked because, it turned out, light did not fit the Newtonian framework.

In a one-two blow, intuitive notions of space, time, and reality were shattered, first by relativity, where space and time vary with perspective, then more seriously by quantum mechanics, where unobserved events dissolve into waves of alternatives. Although correctly describing everyday mechanics as well as such important features of the world as the stability of atoms and the finiteness of heat radiation, the new theories were so offensive to common sense, in concept and consequences, that they inspire persistent misunderstandings and bitter attacks to this day. The insult will get worse. General relativity, superbly accurate at large scales and masses, has not yet been reconciled with quantum mechanics, itself superbly accurate at tiny scales and huge energy concentrations. Incomplete attempts to unite them in a single theory hint at possibilities that exceed even their individual strangeness.

The strangeness begins just beyond the edges of the everyday world. When an object travels from one place to another, common sense insists that it does so on a definite, unique trajectory. Not so, says quantum mechanics. A particle in unobserved transit goes every possible way simultaneously until it is observed again. The indefiniteness of the trajectory manifests itself in the kind of interference pattern created by waves that spread and recombine, adding where they meet in step and canceling where out of step. A photon, a neutron, or even a whole atom sent to a row of detectors via a screen with two slits, will always miss certain detectors, where the wave of its possible positions, having passed through both slits, cancels.

Experimental results forced the quantum view of the world on reluctant physicists piecemeal during the first quarter of the twentieth century and it still has ragged edges. The theory is neat in describing the unobserved, where, for instance, a particle spreads like a wave. It fails to define or pinpoint the act of observation, when the "wave function" collapses and the particle appears in exactly one of its possible places, with a probability given by the intensity of its wave there. It may be when the detector responds, or when the instrumentation connected to the detector registers, or when the experimenter notes the instrument readings, or even when the world reads about the result in physics journals!

In principle, if not practice, the point of collapse can be pinpointed: before collapse, possibilities interfere like waves, creating interference patterns; after collapse, possibilities simply add in a common-sense way. Very small objects, like neutrons traveling through slits, make visible interference patterns. Unfortunately, large, messy objects like particle detectors or observing physicists would produce interference patterns much, much finer than atoms, indistinguishable from common-sense probability distributions because they are so easily blurred by thermal jiggling.

Because, for humans, common sense is easier than quantum theory, workaday physicists take collapse to happen as soon as possible—for instance, when a particle first encounters its detector. But this "early collapse" view can have peculiar implications. It implies that the wave function can be repeatedly collapsed and uncollapsed in subtle experiments that allow measurements to be undone through deliberate cancellation at the experimenter's whim.

This wave function yo-yo is less problematical if one assumes that the collapse happens further downstream where it is more difficult to undo the measurement. Just where the hope of reversal ends is a moving target, as quantum experiments become ever more controlled and subtle. Einstein was troubled by the implications of quantum mechanics, and he devised thought experiments with outcomes so counterintuitive he felt they discredited the theory. Those counterintuitive outcomes are now observed in laboratories and utilized in experimental quantum computers and cryptographic signaling systems. Soon, more advanced quantum computers will allow the results of entire long computations to be undone.

Common sense screams that measurements are real when they register in the experimenter's consciousness. This thinking has led some philosophically inclined physicists to suggest that consciousness itself is the mysterious wave-collapsing process that quantum theory fails to identify. But even consciousness is insufficient to cause collapse in the thought experiment known as "Wigner's Friend." Like the more famous "Schrödinger's Cat," Wigner's friend is sealed in a perfectly isolating enclosure with a physics experiment that has two possible outcomes. The friend observes the experiment and notes the outcome mentally. Outside the leakproof enclosure, Wigner can only describe his friend's mental state as the superposition of the alternatives. In principle these alternatives should interfere, so that when the enclosure is opened one or another outcome may be favored, depending on the precise time of opening. Wigner might then conclude that his own consciousness triggered the collapse when the enclosure was opened, but his friend's earlier observation had left it uncollapsed.

Assuming that effects behave quantum mechanically until some point when their wave functions become so entangled with the world that they are beyond hope of reversal, at which point they behave commonsensically, eliminates philosophical problems for most laboratory physicists. It creates problems for cosmologists, whose scope is the entire universe, for it implies the world is peppered with collapsed wave functions surrounding observing devices. These collapses have no theory and cannot be experimentally quantified and thus make it impossible to set up equations for the universe overall. Instead, cosmologists assume the entire universe behaves as a giant wave function that evolves according to quantum theory and never collapses. But how can a "universal wave function," in which every particle forever spreads like a wave, be reconciled with individual experiences of finding particles in particular positions?

Many Worlds

In a 1957 Ph.D. thesis, Hugh Everett gave a new answer to that question.[19] Given a universally evolving wave function, where the configuration of a measuring apparatus, no less than of a particle, spreads wavelike through its space of possibilities, he showed that if two instruments recorded the same event, the overall wave function had maximum magnitude for situations where the records concurred and canceled where they disagreed. Thus, a peak in the combined wave represents a possibility where, for instance, an instrument, an experimenter's memory, and the marks in a notebook agree on where a particle alighted—eminent common sense. But the whole wave function contains many such peaks, each representing a consensus on a different outcome. Everett had shown that quantum mechanics, stripped of problematical collapsing wave functions, still predicts common-sense worlds—only many, many of them, all slightly different. The "no-collapse" view became known as the "many-worlds" interpretation of quantum mechanics. Its implication that each observation branched the world into something like 10^{100} separate experiences seemed so extravagantly insulting to common sense that it was passionately rejected by many. Although cosmologists worked with the universal wave function, its connection to the everyday world was ignored for another twenty years.

Recent subtle experiments confirming the most mind-bending predictions of quantum mechanics, including the development of quantum computers, have lifted many-worlds' stock relative to traditional interpretations that require influences to leap wildly across time and space to explain the observed correlations. The theoretical trail pioneered by Everett is becoming traveled and extended. Since the late 1980s James Hartle and Murray Gell-Mann have investigated its underlying notions of measurement and probability.

Everett had demonstrated that the conventional rules for collapsing the wave function to measurement-outcome probabilities from "outside" a system were consistent with what would be reported by (each version of) the uncollapsed observer "inside," thus removing the requirement for an outside or a collapse and raising our consciousness to existence of many worlds. He made no attempt to show how those peculiar measurement rules arose in the first place. Gell-Mann and Hartle are asking this difficult question. They are far from a final resolution, but their work so far shows just how special—or illusory—the common-sense world really is.

Hartle and Gell-Mann note that if we were to try to observe and remember events at the finest possible detail—around 10^{-30} centimeters, far smaller than anything reachable today—the interference of all possible worlds would present a seething chaos with no permanent structures, no quiet place to store memories, effectively no consistent time. At a coarser viewing scale—10^{-15} centimeters, the submicroscopic world touched by today's high-energy physics—much of the chaos goes unobserved, and multiple worlds merge together, canceling the wildest possibilities, leaving those where particles can exhibit a consistent existence and motion, if still jaggedly unpredictable, through a vacuum that boils with ephemeral "virtual" energy. Everyday objects have the smooth, predictable trajectories of common sense only because our dim senses are coarser still, registering nothing finer than 10^{-5} centimeters. At scales larger than the everyday (or the Hartle–Gell-Mann analysis), the events we consider interesting are blurred to invisibility, and the universe is increasingly boring and predictable. At the largest possible scale, the universe's matter is canceled by the negative energy in its gravitational fields (which strengthen while releasing energy, as matter falls together), and in sum there is nothing at all.

No complete theory yet explains our existence and experiences, but there are hints. Tiny universes simulated in today's computers are often characterized by adjustable rules governing the interaction of neighboring regions. If the interactions are made very weak, the simulations quickly freeze to bland uniformity; if they are very strong, the simulated space may seethe intensely in a chaotic boil. Between the extremes is a narrow "edge of chaos" with enough action to form interesting structures, and enough peace to let them persist and interact. Often such borderline universes can contain structures that use stored information to construct other things, including perfect or imperfect copies of themselves, thus supporting Darwinian evolution of complexity. If physics itself offers a spectrum of interaction intensities, it is no surprise that we find ourselves operating at the liquid boundary of chaos, for we could not function, nor have evolved, in motionless ice nor formless fire.

The odd thing about the Hartle–Gell-Mann spectrum is that it is not some external knob that controls the interaction intensity, but varying interpretations of a single underlying reality made by observers who are part of the interpretation. It is, in fact, the same kind of self-interpretation loop we encountered when considering observers inside simulations. We are who we are, in the world we experience, because we see ourselves that way. There are almost certainly other observers in exactly the same regions of the wave function who see things entirely differently, to whom we are simply meaningless noise.

The similarity between Everett's "many worlds" and the philosophical "possible worlds" may become stronger yet. In "many worlds" quantum mechanics, physical constants, among other things, have fixed values. Gravity, in objects like black holes, loosens the rules, and a full quantum theory of gravity may predict possible worlds far exceeding Everett's range—and who knows what potent subtleties lie even further on? It may turn out, as we claw our way out through onion layers of interpretation, that physics places fewer and fewer constraints on the nature of things. The regularities we observe may be merely a self-reflection: we must perceive the world as compatible with our own existence—with a strong arrow of time, dependable probabilities, where complexity can evolve and persist, where experiences can accumulate in reliable memories, and the results of actions are predictable. Our mind children, able to manipulate their own substance and structure at the finest levels, will probably greatly transcend our narrow notions of what is.

Questioning Reality

Like organisms evolved in gentle tide pools, who migrate to freezing oceans or steaming jungles by developing metabolisms, mechanisms, and behaviors workable in those harsher and vaster environments, our descendants may develop means to venture far from the comfortable realms we consider reality into arbitrarily strange volumes of the all-possible library. Their techniques will be as meaningless to us as bicycles are to fish, but perhaps we can stretch our common-sense-hobbled imaginations enough to peer a short distance into this odd territory.

Physical quantities like the speed of light, the attraction of electric charges, and the strength of gravity are, for us, the unchanging foundation on which everything is built. But if our existence is a product of self-interpretation in the space of all possible worlds, this stability may simply reflect the delicacy of our own construction—our biochemistry malfunctions in worlds where the physical constants vary, and we would cease to be there. Thus, we always find ourselves in a world where the constants are just what is needed to keep us functioning. For the same reason, we find the rules have held steady over a long period, so evolution could accumulate our many intricate, interlocking internal mechanisms.

Our engineered descendants will be more flexible. Perhaps mind-hosting bodies can be constructed that are adjustable for small changes in, say, the speed of light. An individual who installed itself in such a body, and then adjusted it for a slightly higher lightspeed, should then find itself in a physical universe appropriately altered, since it could then exist in no other. It would be a one-way trip. Acquaintances in old-style bodies would be seen to die—among fireworks everywhere, as formerly stable atoms and compounds disintegrated. Turning the tuning knob back would not restore the lost continuity of life and substance. Back in the old universe everything would be normal, only the acquaintances would witness an odd "suicide by tuning knob." Such irreversible partings of the way occur elsewhere in physics. The many-worlds interpretation calls for them, subtly, at every recorded observation. General relativity offers dramatic "event horizons": an observer falling into a black hole sees a previously inaccessible universe ahead at the instant she permanently loses the ability to signal friends left outside.

Visiting offbeat worlds, where the dependable predictability of the common sense no longer holds, is probably much too tricky for crude techniques like the last paragraph's knob turning. It must be far more likely that mechanical fluctuations or other effects persistently frustrate attempts to retune a body than for physical constants to actually change. Yet once our descendants achieve fine-grain mastery of extensive regions of the universe, they may be able to orchestrate the delicate adjustments needed to navigate deliberately among the possibilities, perhaps into difficult but potent regions shaped by interrelationships richer than those of matter, space, and time. Time travel, a technology faintly visible on our horizon, may mark merely the first and most pedestrian route in this limitless space.

Until Death Do Us Part

We can't yet leave the physical world in chosen directions, but we are scheduled to leave it soon enough in an uncontrolled way when we die. But why do we seem so firmly locked to the simple physical laws of the material world before death? This is a most fundamental question if one accepts that all possible worlds are equally real. Artificial intelligence programs, which recreate the psychological state of nervous systems without simulating the detailed physical substance that underlies them, and virtual realities, which allow unphysical magical effects like teleportation, suggest that our own consciousnesses can exist in many possible worlds that do not follow our physical laws. This question of why our universe seems so firmly yoked to physical law has hardly been asked in a scientific way, let alone answered. But the answer may be related to Einstein's observation that mathematics seems to be unreasonably effective in describing the physical world. This unreasonableness shows itself in the plausible, already partially fulfilled, quest of physics for a "Theory of Everything," perhaps a simple differential equation whose solution implies our whole physical universe and everything in it!

In our daily meanders, we are more likely to stumble across a particular small number (say "5") than a particular large one (say "53783425456"). The larger number requires far more digits to simultaneously fall into place just so, and thus is far less likely. Similarly, although we exist in many of all possible universes, we are most likely to find ourselves in the simplest of those, the few that require the least number of things to be just so. The universe's great size and age, its physical laws, and our own long evolution may be just the working of the simplest possible rules that produce our minds.

Our consciousness now finds itself dependent on the operation of trillions of cells tuned exquisitely to the physical laws into which we evolved. It continues from moment to moment most simply if those laws continue to operate as they have in the past. Thus, with overwhelming probability, we find the laws are stable. In the space of all possible universes, we are bound to the same old one. As long as we remain alive.

When we die, the rules surely change. As our brains and bodies cease to function in the normal way, it takes greater and greater contrivances and coincidences to explain continuing consciousness by their operation. We lose our ties to physical reality, but, in the space of all possible worlds, that cannot be the end. Our consciousness continues to exist in some of those, and we will always find ourselves in worlds where we exist and never in ones where we don't. The nature of the next simplest world that can host us, after we abandon physical law, I cannot guess. Does physical reality simply loosen just enough to allow our consciousness to continue? Do we find ourselves in a new body, or no body? It probably depends more on the details of our own consciousness than did the original physical life. Perhaps we are most likely to find ourselves reconstituted in the minds of superintelligent successors, or perhaps in dreamlike worlds (or AI programs) where psychological rather than physical rules dominate. Our mind children will probably be able to navigate the alternatives with increasing facility. For us, now, barely conscious, it remains a leap in the dark. Shakespeare's words, in Hamlet's famous soliloquy, still apply:

> *To die, to sleep;*
> *To sleep: perchance to dream: ay, there's the rub;*
> *For in that sleep of death what dreams may come*
> *When we have shuffled off this mortal coil,*
> *Must give us pause: there's the respect*
> *That makes calamity of so long life;*
> *For who would bear the whips and scorns of time,*
> *The oppressor's wrong, the proud man's contumely,*
> *The pangs of despised love, the law's delay,*
> *The insolence of office and the spurns*
> *That patient merit of the unworthy takes,*
> *When he himself might his quietus make*
> *With a bare bodkin? who would fardels bear,*
> *To grunt and sweat under a weary life,*
> *But that the dread of something after death,*
> *The undiscover'd country from whose bourn*
> *No traveller returns, puzzles the will*
> *And makes us rather bear those ills we have*
> *Than fly to others that we know not of?*
> *Thus conscience does make cowards of us all;*
> *And thus the native hue of resolution*
> *Is sicklied o'er with the pale cast of thought,*
> *And enterprises of great pith and moment*
> *With this regard their currents turn awry,*
> *And lose the name of action.*

Notes

1. Stephen Jay Gould, *Time's Arrow, Time's Cycle: Myth and Metaphor in the Discovery of Geological Time* (Jerusalem-Harvard Lectures), Harvard, 1987.

2. Bob Connolly and Robin Anderson, *First contact* (motion picture) produced in association with The Institute of Papua New Guinea Studies, New York: Filmmakers Library, 1984.

3. Richard Dawkins, *The Selfish Gene,* 2nd edition, Oxford, 1989.

4. Robin I. M. Dunbar, *Grooming, Gossip, and the Evolution of Language,* Harvard 1997.

5. Norbert Wiener, *Cybernetics: Or Control and Communication in Animal and the Machine,* 2nd edition, MIT, 1965.

6. W. Grey Walter, *The Living Brain,* Norton, 1963.

7. Bertram Raphael, *The Thinking Computer: Mind inside Matter,* Freeman , 1976.

8. DARPA was formed (as ARPA) in the aftermath of the unexpected 1957 launch of the Soviet Sputnik, the first satellite, to fund research and forestall other such "technological surprises." It was the major funder for Artificial Intelligence in the 1960s and 1970s and remains a significant source.

9. D. J. Goldhaber-Gordon, M. S. Montemerlo, J. C. Love, G. J. Opiteck, and J. C. Ellenbogen, *Overview of Nanoelectronic Devices,* Proceedings of the IEEE, April 1997.

10. In October and November 1996, William McCune's program EQP found two proofs for the "Robbins Conjecture" that certain axioms define a Boolean algebra. The program built on twenty-five years of computer theorem-proving work at Argonne under Larry Wos. The result is detailed in: William McCune, "Solution of the Robbins Problem," *Journal of Automated Reasoning*, 1997.

11. General relativity, as formulated by Einstein, is a set of differential equations describing gravity that are difficult to solve because they are nonlinear. The nonlinearity has also prevented them from being cast in the framework of quantum mechanics in the same way the linear equations of special relativity and electromagnetism were transformed into the ultra-successful theory of quantum electrodynamics. In 1989 Abhay Ashtekar discovered a change of variables that linearized Einstein's equations and did permit a quantum mechanical version. Unexpectedly, the combination implied a strange topology in the resulting spacetime. Instead of general relativity's smooth continuum, the quantized theory described a spacetime made of loops at the Planck scale of 10^{-33} centimeters. Subsequent work showed this "loop variables" theory implied that area and volume were quantized at the Planck scale. In the 1990s, the theory was linked with the mathematical theory of knots, as well as with superstring theories of the other fundamental forces. More and more, the fabric of spacetime is beginning to look like literal cloth, though of a very, very fine and very, very strange kind!

12. No reaction at energies achievable today can simply convert a single atom into photons—the total number of protons or neutrons remains unchanged. It is possible, however, to convert an atom to photons by combining it with a twin atom of antimatter. Antiprotons and antineutrons count as negative protons and neutrons, so their total number in a matter/antimatter pair is zero and thus unchanged before and after the mutual annihilation. It *is* possible to change the number of neutrons and protons at much higher energies. According to a theorem by Stephen Hawking, a black hole is like a

giant elementary particle, described solely by its mass, spin, and charge. Protons or neutrons swallowed by a black hole lose their identity. When the black hole later emits their corresponding energy, it is in the form of thermal Hawking radiation, which for larger black holes, is mostly composed of photons. An advanced technology may be able to orchestrate such transformations in a more controlled manner.

13. Jacob Bekenstein and Marcelo Schiffer, "Quantum Limitations on the Storage and Transmission of Information," *International Journal of Modern Physics* v1, pp. 355–422, 1990.

14. The recreations may be imperfect and uncertain at first, though much, much better than any historical drama or simulation we can stage today. The fidelity will improve as observed clues and mental inference power mount. Even with uncertainty, high fidelity can be achieved by recreating the whole range of possibilities allowed by the uncertainty—in fact such "subjunctive" simulations can be part of the inference process that weeds out possible worlds. Any simulations that produce results that contradict physical observations, however minor, can be eliminated as possibilities. As more and more is deduced about the past world, the range of possibilities that must be simulated shrinks, and the job of simulation becomes easier.

15. Kip S. Thorne, *Black Holes and Time Warps: Einstein's Outrageous Legacy*, W. W. Norton, 1995.

16. The one glaring exception is the ugly "wave-function-collapse" hypothesis traditionally invoked to explain the process of measurement in quantum mechanics. Between measurements, the wave function that encodes all the possible states of a system, and their mutual interferences, evolves in a deterministic time-reversible manner. When a measurement is made, however, the wave function seems to "collapse," and just one of the possibilities is actually observed. Just how and exactly when the collapse actually happens has been a matter of controversy since the idea was proposed. There is no

mathematics to pinpoint the collapse, and no ad hoc placement of the event explains the result of every possible experiment. In fact, for cosmologists who consider the wave function of the entire universe, any collapse anywhere, anytime in the universe is untenable, because it messes up the universe's evolution. In 1957 Hugh Everett III showed how a forever-uncollapsed wave function merely *seems* to collapse for particular observers, in one elegant step eliminating all of collapse's problems. Despite its overwhelming advantages in unambiguously explaining the observations of every experiment, a generation of physicists rejected Everett's "relative state" interpretation because they found its implications too disturbing to their intuitions. It is rapidly gaining ground, however, in a newer generation. Everett's uncollapsed wave function is fully time reversible. This issue is discussed further in Chapter 7.

17. David Deutsch, *The Fabric of Reality: The Science of Parallel Universes and Its Implications*, Allen Lane, 1997. Deutsch's resolution of time-travel paradoxes is somewhat different from the one I opt for in this chapter. In his, time travelers who make round trips to the past return to find themselves in different alternate universes, with memories about recent events in disagreement with those of nontravelers. Although I don't explicitly consider physical time travelers, my assumptions possibly would favor a model in which consistency is enforced by wave-function cancellation of inconsistencies, where the traveler always returns to the same universe, albeit one made strange by the time loop. I'm not certain if this necessarily follows, though. It depends on just how brain function and its interpretation as trains of consciousness interacts with time travel. To me, there is still enough uncertainty about this matter to allow several possibilities. Perhaps these are not even mutually exclusive, but depend on just how the time travel is done. For instance, a trip by wormhole may produce a different effect on the traveler than one by matter transmitter over a tachyon relay because they involve different topologies of

the surrounding spacetime. It's fun, fun, fun in these wild, unexplored frontier areas!

18. The example of a rock interpreted as a mind comes from an appendix to Hilary Putnam's book *Representation and Reality*, Bradford Books, MIT Press, 1991. Putnam notes that any system with nonrepeating states has an interpretation as any finite state machine, and thus that finite state machines, or computations, cannot alone define or create a mind. In making this argument, Putnam implicitly rejects as absurd the possibility that everything contains a mind. "Obviously" minds reside only in certain special objects, like people and some animals. Unfortunately, the more carefully one thinks about the distinction, the more elusive it gets, especially in a world of interactive machines, intelligence movingly encoded into books, and creatures like whales, who may have wonderful minds separated from ours by a language barrier. The fundamental problem in philosophy of mind seems to be that our intrinsic intuitions about mind and matter are inconsistent. As with the new theories of physics and biology, one probably must break with intuition to get a consistent model of mind. This chapter makes an unintuitive leap, that minds and worlds don't exist objectively at all, but, like beauty and value, exist only in a context, as interpretations in the eye of a beholder. Of course, beholders exist only in minds and worlds, so we live in a circular illusion and there is no such thing as context-free objective existence. We don't really exist at all; we just think we do. But so do self-beholders encoded in rocks.

    A reviewer objected to the example on the grounds that rocks are usually in thermal equilibrium, and thus lack the energy flow to drive organized computation. Actually, interpretations can as easily be made of reversible as of irreversible processes. For instance, the equilibrium rotation and orbits of the earth are often used as a clock. A rock's atomic motions are like planetary orbits, just more subtle and convoluted. Interpretations can even be made where there is no motion at all. A sequence of positions on the surface of the rock, for instance, can be mapped to the states of a computation. To make the point less vacuously, consider logarithms, less controversial than minds, but no less a matter of contextual interpretation. A scientific calculator can be interpreted as having logarithms. Press some digit keys and the "log" button and a pattern appears on the display that you (but not your three year old) interprets as a logarithm. Your calculator, like a conventional simulation, is powered by electricity, and is not in equilibrium. But a printed piece of paper can be in equilibrium and yet still be interpreted as having logarithms. That's how we did logarithms in the old days—by finding them in a log table lying on the desk as inertly as any rock.

19. Hugh Everett, *Many-Worlds of Interpretation of Quantum Mechanics*, Princeton, 1973.

Acknowledgments

I can't acknowledge everyone who has influenced this book—the trail is decades long, and many ideas began as partly understood suggestions that grew, sometimes toward the original intent, often elsewhere. Some of the most obvious credits are sprinkled throughout the text. In recent years I've found it rewarding to air emerging ideas to the interesting people found in some of the internet's virtual coffee shops. This book has benefited from extended discussions in the lairs of the extropians, comp.ai, comp.robotics, sci.space, sci.physics, sci.math, and other groups. It has, of course, also benefited from old-style interactions in my community at Carnegie Mellon University and elsewhere.

This book, as my previous one, was typeset by Mike Blackwell in Don Knuth's TeX language. The illustrations were created or edited with Adobe Photoshop and Deneba Canvas on a G3 Macintosh, whose 500 MIPS, 128 MB RAM, 4 GB disk, and 100 Mbps ethernet give it exactly 1,000 times the capacity of the 0.5 MIPS, 128 KB original Mac (and accompanying box of 400 KB floppies) that started my previous book in 1985. A bit behind the curve, the G3's two bulky CRT monitors are sated by just 340 times the data that fed the original Mac's tiny screen.

The figure on page 6 is assembled from clip art purchased on line from ArtToday and PhotoDisc. Small clip art images in subsequent pages were obtained from ArtToday or taken from public web pages. Much of the information sprinkled throughout the book, in fact, was compiled from intensive web browsing. Some of the relevant links can be found on the web page whose URL is in the Preface. Page 16's illustration is a composite of photographs provided by Bruce Baumgart, Rod Brooks, and Les Earnest. The one on page 18 is derived from Richard Pawson's *The Robot Book*, Windward, 1985, p. 14. Page 19's comes from the Johns Hopkins University Applied Physics Laboratory. Page 27's is from SRI International. Martin Frost restored the 1979 Cart's eye image on page 38 from ancient computer backup tapes. Mary Jo Dowling provided the photo for page 42. Richard Wallace's 1994 New York University computer architecture class, especially students Mohammed Kadir, Irina Pirotskaya, Alexandr Shenker, and Scott Sterling supplied 1987 to 1994 data for the graph on page 60. The cartoon images on pages 89 and 90 were drawn by Hans-Jürgen Marhenke for a translation by Thomas Schult that appeared in June 1996 in the German **C'T** magazine. The 3D robot designs on pages 94, 105, 108, 116, and 124 were generated by Jesse Easudes using the Pro-Engineer program on Silicon Graphics workstations. The figure on page 99 is a composite of images by Honda Motors Corp. and Ella Moravec. The figure on page 151 is derived from an illustration by Richard Ellis found in his book *Deep Atlantic*, Knopf, 1996, p. 151. Marc LeBrun and Kevin Dowling alerted me to this and other pictures of basket stars. The 3D images on pages 152, 153, 160, 181, 184, 186, and 188 were generated in the VRML language and rendered in Netscape with CosmoPlayer plug-in on Silicon Graphics workstations. Mike Blackwell and Martin Martin provided technical assistance. Many thanks to them, and to all the others.

Index

Adam 127
Advanced Research Projects Agency 41
agape 119
aggression 120, 142
aging population 48
AGV 47
AI Lab 16
Albus, James 133
alien intelligence 67
ALV 41–2
ALVINN 43–4, 102
amacrine cell 53
Amazon 3
America 9, 17
American Indian 127
Amish 8, 141
analog computer 86
Analog magazine vii
Analytical Engine 84
Anderson, Robin 213
Anglican church 75
animating principle 111
Anthropic Cosmological Principle 201
anti trust 140
Apollo 25
Apple Computer 52
Applied Physics Laboratory 18, 220
Arabia 135
Arastradero Road 15
Archimedes 163
Argonne National Laboratory 65, 214
argument from consciousness 74, 82–3
argument from continuity in the nervous system 74, 85–6
argument from extrasensory perception 74, 87–8
argument from informality of behavior 74, 86–7
arguments from various disabilities 74, 84
Aristotle 78
ARPA 41
artificial heart 142
artificial human 93
Artificial Intelligence Project 15
artificial limb 142
artificial reality 72
Ashtekar, Abhay 214
Asia 9
Asimov, Isaac 140
assembly language 28
atom bomb 20
AT&T 66–7, 71
Austin 23
Australia 2, 133
Automatic Guided Vehicles 47
Autonomous Land Vehicles 41

Babbage, Charles 84–5
baby boom 4, 134
Bacon, Roger 12, 163
bad air 3
Barrow, John 201–2
basket starfish 150–1
Baumgart, Bruce 29, 219
BBC 20
Beast 19
Bekenstein, Jacob 166, 215
Bell Labs 67
Belle 67, 71
Bem, Daryl 88
Berkeley, University of California 121
Berlin 37
Bernstein, Alex 66, 71
big bang 201
bipolar cell 53
black hole 175, 207–9, 214
Blackwell, Mike 219–20
blocks world 28
Boeing 777 2
Boston 36
bottom up 108
brain in a vat 169, 191–2
Brazil 3
Britain 17–20, 72, 75, 132
Brooks, Rod 46–7, 220

buckytube 157–9
Bundeswehr University 41
buried wire 16, 47
business plan viii
Byron, George Gordon 84

Caltech 173
camera solver 30
Cantor, Georg 81
Čapek, Karel 93
Carnegie Mellon University 5, 41–2, 53, 66, 219
Cart 16, 25, 28–34, 37–8
CAUTION ROBOT VEHICLE 15
CDC 66, 71
chess 17, 21–2, 54, 66–72, 85, 125, 154, 171, 186–9
Chess x.0 71
chessboard 17, 67
Chevy van 41
China 4
Chinook 66
Chiptest 71
Christian 119
chronology protection conjecture 179
cleaning robot 92–3, 97–8
CM-5 36
CMU 5, 53, 219
code breaking 17, 73
Cog 46–7
cold war 131
Colossus 20
combinatorial explosion 23
commercial robot 92–3
comp.ai 219
comp.robotics 219
computer engineering 1
computer science department 5
computer vision vii, 29, 45
Computer-Controlled Cars 16, 26
Computing Machinery and Intelligence 73–88
conditioning module 100–6, 112–14, 118–20
Connecticut 48
Connolly, Bob 213
Consciousness Explained 122
constitution 146–7, 155
construction set vii
Contact 173
Control Data Corporation 66, 71
Cope, David 66
Cornell University 88
corporate law 139–40, 154–5
corporate tax 139–41
cosmic censorship 175
Cray 66, 71
Cray Blitz 71
C'T magazine 220
customer service 138
Cybermotion 48
cybernetics 17-19, 23
cyberspace 164–8, 171–2
Cyc 23, 103, 109
Czech 93

da Vinci, Leonardo 12, 163
Daedalus 163
Daimler-Benz 37
dangerous robot 97
DARPA 41, 213
Dartmouth College 20
Darwin, Charles 75, 81, 104, 121, 146–7, 165, 207
Dawkins, Richard 4, 213
DDT 56
de Chardin, Tielhard 202
death 20, 136, 209–11
DEC 26, 65
Deep Blue 21, 54, 66–72, 187
Deep Thought 66
Defense Advanced Research Projects Agency 41
delivery robot 92, 98
Dennett, Daniel 122
Denning Mobile Robotics 32–3
Denver 41
Department of Defense 41
Descartes, René 78, 121, 191
Deutsch, David 180, 216
developed world 1
devil 146, 191, 198
diagonalization 81
Digital Equipment Corporation 26, 65
Dirac, Paul 161–2
Disney, Walt 105
DNA 125, 147–8
docking station 94
DOD 41
Dowling, Kevin 220
Dowling, Mary Jo 220
dream 145, 169, 210–11
Dreyfus, Hubert 121
driver's assistant 45
Duke University 87
Dunbar, Robin 4–5, 213

Earnest, Les 220
Easudes, Jesse 220
economy of scale 38

edge of chaos 207
Edinburgh University 88
Einstein, Albert 174–5, 210, 214
electrochemical 56, 194–5
electromechanics 47, 57–61
Elements 110
Elfes, Alberto 32
Ellenbogen, J. C. 213
Ellis, Richard 220
Elsie 18
EMI 66, 69
England 18, 75, 84
EQP 65, 69, 214
escape clause 143
escape velocity 2–4
ESP 74, 87
Euclid 78–80, 110
Europe 9, 131
event horizon 175, 202, 209
Everett III, Hugh 205–7, 216–18
Ex 144–50, 154–5, 160–7, 173, 180
extrasensory perception 74, 87
extraterrestrial 15
extropians 219

factory vehicle viii
Faust 127, 146
Feynman, Richard 177
first generation 95–8, 112, 116
Fischer, Bobby 67
fish eye 37
five hour 16, 31
flood 70–72, 130
fourth generation 108–10, 113–15, 118–20, 123–5, 140
Franklin, Benjamin 163
French Swiss 136
Frost, Martin 220
Fuller, Buckminster 157

Galileo Galilei 75
Gelernter, Herbert 109–10
Gell-Mann, Murray 206–7
general purpose 25
general relativity 161
General Telephone and Electric 15
genetic algorithm 104
Gennery, Donald 30
genome 126
German Swiss 136
Germany 20, 37, 41, 199, 220
ghost in the machine 121
giant brain 20–1, 70, 84
Gnu C 36
God 14, 75–7, 121, 167
Goddard, Robert 13
Gödel, Kurt 78–82, 174
Goldhaber-Gordon, D. J. 213
Gort 15
Gould, Stephen Jay 213
graduate student 29–31
grandmaster 21, 67–9, 85
Greek 119
Grey Walter, William 18, 46, 213
GTE 15
guard robot 93

Hamlet 211
hand code 55
hand eye 48
Hartle, James 206–7
Hawking, Stephen 179, 214–15
heads in the sand objection 73, 77–8
heat sink 61
Heathkit 45
Hero 45
highway of the future 45
Hilbert, David 78–9, 93
Hitech 71
Holmes, Sherlock 167
home robot viii, 92
homosexuality 136
Honda Motors 99, 220
Honorton, Charles 88
Hopkins Beast 18, 46
horizontal cell 53
housecleaning 96–8
human reasoning ix
hunter gatherer 3, 9, 127, 135

IBM 64–6, 71, 85
IBM 704 64–6
ice age 3
ice cream 53
Indian 127
industrial revolution 9, 128, 131
industrial robot 25
inertial navigation 46
Intel 66
interaction model 103–7
interconnection 43
interpretation 165, 193–202
Islam 135
Italian 136

Jacquard, Joseph-Marie 84
Japan 40, 48, 99
Jefferson, Geoffrey 82
Jesuit 202
Jet Propulsion Laboratory 25
Jochem, Todd 44–5

JOHNNIAC 20
Johns Hopkins University 18, 220

Kadir, Mohammed 220
Kasparov, Garry 21–2, 54, 66–71
Kerr, Roy 175
KL-10 31
Knuth, Don 219
Kodak 3
Kopenawa, Davi 3, 8, 135
Kuwait 135

Lady Lovelace's objection 74, 84–5
land rush 163
landscape of human competence 70
language translation 64
language understanding 109
laser 27, 37, 41, 45, 177, 180–3
Laws of Robotics 140
LeBrun, Marc 220
Lederberg, Joshua 28
Leonardo da Vinci 12, 163
life expectancy 135
Life magazine 26
lightning calculators 22
Linares 67
Lister Oration 82
lizard viii, 58–60, 68, 71, 92, 95
Logic Theorist 20
Love, J. C. 213
Lovelace, Ada Augusta 74, 84–5

Macintosh 56–8
MackHack 71
Magellanic Cloud 174
Manchester University 72–3, 82
manipulator 21, 25, 31, 45–7, 83, 95–7, 104–5, 148–151
many-worlds interpretation 205–9, 216
Marhenke, Hans-Jürgen 220
Mariner 28–9
Mario Brothers 199
Mars 25, 28–30, 65
Martin, Martin 38, 220
Martin-Marietta Corporation 41
mass production 25, 37
Massachusetts Institute of Technology 21, 26, 46
mathematical objection 74, 78–82
matter transmitter 216
Matthies, Larry 32
Maxwell, James Clerk 177
MCC 23
McCarthy, John vii, 15–16, 20–1, 26–9, 40
McCune, William 214
mechanical engineering 47–8
Mechanical Engineering Laboratory 40
Microelectronics and Computer Consortium 23
microscopic finger 10
mid brain 110
mind children viii, 13, 77, 125–6, 163, 199, 208, 211
mind fire 14, 191
Mind journal 73
miniaturization 59, 148
Minsky, Marvin 21
MIT 21, 26, 46
mobile robot laboratory 32
molecular simulation 2
monkey viii, 4, 58–60, 67–8, 71, 104
Montemerlo, Mike 213
Moravec, Ella 220
Moravec, Hans 25, 29–37
Mount Rushmore 199
mouse viii, 58–60, 68, 71, 98
MRI 189
MRL 32
Munich 41
Murphy's Law 184–6
music box 84
music composition 66

naked 2–3
nanotechnology 144
Napoleon Bonaparte 136
NASA 25, 28, 96
National Bureau of Standards 133
Navlab 41–4, 52, 56
Nealey, Robert 66
Near East 4
negative conditioning 101, 120, 123, 140
negative-time-delay 180–4
neurotransmitter 85
New Guinea 2
New York 3, 6
New York University 220
Newell, Allen 20
Newman, Ezra 175
Newton, Isaac 2, 52, 173, 176–8, 202
nine eyes 31
Niven, Larry 179
NMR 63, 189
Nobel prize 28
North America 133–4
North Carolina 87
NP 185–7
nuclear magnetic resonance 63, 189

Oberth, Hermann 13
OCR 25, 52, 130

NP 185–7
nuclear magnetic resonance 63, 189

Oberth, Hermann 13
OCR 25, 52, 130
off road 43
Office of Naval Research 33
oil exploration 66
Omega Point 202
omnidirectional wheel 94, 105
ONR 33
Opiteck, G. J. 213
optic nerve 53
Oxford University 180

P2 and P3 robots 99
Panama 133
paradox 88, 173, 178–84, 216
paranormal 87–8
Pasadena 25
pattern recognition 23
Pawson, Richard 220
PDP computer 26–9, 65
Pennsylvania 8
Penrose, Roger 79, 122
perception model 104, 107
petro kingdom 135
Ph.D. 16, 25, 32, 42–5, 205
Physics of Immortality 201
Pirotskaya, Irina 220
Pittsburgh 3
placental mammal 134
Planck, Max 161, 164, 214
planetary defense 141, 154
Plato 122, 196–8
Plymouth minivan 44
Polanyi, Michael 72
Polaroid 32, 35
Pomerleau, Dean 43–4, 102
positive conditioning 101, 123, 140
possible worlds 194, 200, 207–10, 215
Post, Emily 86
preschool adventures vii
price/performance viii, 60, 65
Principia Mathematica 20
printed circuit 59
probability fuse 183–6
psychic 87–8
psychopath 146
Putnam, Hilary 217

Quam, Lynn 28–9
quantum collapse 122
quantum computer 62–3, 180, 189, 204–6
quantum dot 62
quantum gravity 164
quantum interference logic 62
qubit 63, 189

RALPH 44, 102
rangefinder 26, 37, 41, 45
Raphael, Bertram 213
rational thought 24
reading program 25, 52, 130
reasoning program vii, 24
Reddy, Raj 53
reincarnation 77
resurrection 142, 173
retina 53–4
Rhine, Joseph 87–8
road sign 15, 26–8
Robbins, Herbert 65, 69, 214
robot arm 21, 25, 31, 45–7, 83, 95–7, 104–5, 148–151
robot psychologist 103
robot vacuum viii, 40, 92–4
Rossum's Universal Robots 93
Russell, Bertrand 20
Russia 64, 122
Ryle, Gilbert 121

Sagan, Carl 173
Samuel, Arthur 66
San Diego 45
Saudi Arabia 135
scanning tunneling microscope 151
Schaeffer, Jonathan 66
Scheinman, Victor 31, 47
Schiffer, Marcelo 215
Schmidt, Rod 16, 28
Schrödinger, Erwin 204
Schult, Thomas 220
science fiction 13, 141, 173, 179, 191
sci.math 219
sci.space 219
Searle, John 122
second generation 98–104, 112, 116
Selfish Gene, The 213
sensory deprivation 170
service economy 132
seven-league boot 12
sex 82, 118
Shakespeare, William 211
Shakey 25–8, 107
Shaw, George 20
Shenker, Alexandr 220
SHRDLU 112
Simon, Herbert 20
single-electron transistor 62
six leg 46
Skinnerian conditioning 101

sonar 19, 32–8, 45
Sorcerer's Apprentice 105
South America 133–4
Soviet Union 64
space travel 111, 149
Spain 67
speech recognition 25, 52–5, 65
Sphinx II 53
spider 58–60, 68, 71, 150
spirit 76, 111, 121, 147, 194
sports announcer 115
spot weld 24
SRI International 25–7, 107, 220
Stalinism 122
Stanford Artificial Intelligence Laboratory vii, 15–16, 48, 65, 129
Stanford Cart 16
Stanford Hospital 48, 91
Stanford University 15, 21, 25–8, 53
state of the art 48
Sterling, Scott 220
stone age 3, 7–8
Strategic Computing Initiative 41
STRIPS 107–8
superheterodyne 146
superposition 62–3, 189
superrationality 124–5
supersymmetry 161
Switzerland 136–7

tachyon 174, 179–80, 216
technological surprise 64–5, 213
teleportation 209
telepresence 168–9
Terragator 42
TeX 219
Texas 23
text recognition 25, 52, 130
The Day the Earth Stood Still 15
The Mind and the Computing Machine 72
The Mind of Mechanical Man 82
theological objection 73–7
theorem proving 20–2, 26, 65, 70–2, 80, 107–9
Theory of Everything 210
Thinking Machines Corporation 36
third eye 121
third generation 104–8, 112–14, 117–18
this time for sure viii
Thompson, Ken 67
Thorne, Kip 173–8, 215
Thorpe, Chuck 32, 42–3
three dimensions 35–40, 52, 92–4, 97, 106, 129, 148

time travel 88, 173–89, 216
Tipler, Frank 175, 201–2
Today's Computers, Intelligent Machines and Our Future vii
top down 108
Transitions Research Company 48
traveling salesman 185-187
TRC 48
tree of knowledge 143
triangulation 29, 38
Tsiolkovsky, Konstantin 13
Turing, Alan 20, 72–88, 93, 199
Turing machine 79–81
Turing test 73–88
TV 15, 29–31, 37, 40, 45
two dimensions 35–8, 45, 92

U boat 20, 199
uncomputable number 80–1
United Arab Emirates 135
United States 65, 128, 131, 134
universal computer 20, 58
universal robot 25, 82–3, 91–126
universal Turing machine 79–80, 93
University of Alberta 66
unsolvable problem 20
URL ix
Usenet 137–8
utility robot viii, 92-5

van 42–3, 52, 58
Vax 65
Verne, Jules 12, 163
Victorian 84, 173, 202
Virginia 48
virtual reality 168–9, 191–2
von Neumann, John 20
Voyager 167

wall plug 19
Wallace, Richard 220
washing machine 98
Washington, D.C. 3, 45
wave function collapse 203–6, 215–16
weather control 155–6
Wells, Herbert G. 12, 173
wet lab 2
Wheeler, John 177
whistling-in-the-dark 84
Whitehead, Alfred N. 20
wide beam 32
Wiener, Norbert 17, 213
Wigner, Eugene P. 204
Winograd ,Terry 112
wolf pack 20
World War I 136

World War II 17, 20, 130, 134, 155
World Wide Web 93, 138
wormhole 173, 176–80, 216
Wos, Larry 214

Yanomami 3–5, 135

Zen 115
zeroth generation 95
zombie 197, 201